MONOGRAPHS ON
STATISTICS AND APPLIED PROBABILITY

General Editors

D.R. Cox, D.V. Hinkley, N. Keiding, N. Reid, D.B. Rubin and B.W. Silverman

(Full details concerning this series are available from the Publishers).

Cyclic and Computer Generated Designs

SECOND EDITION

J.A. JOHN

Centre for Applied Statistics
University of Waikato
Hamilton
New Zealand

AND

E.R. WILLIAMS

CSIRO Division of Forestry
Canberra
Australia

CHAPMAN & HALL

London · Glasgow · Weinheim · New York · Tokyo · Melbourne · Madras

Published by Chapman & Hall, 2-6 Boundary Row, London SE1 8HN, UK

Chapman & Hall, 2-6 Boundary Row, London SE1 8HN, UK

Blackie Academic & Professional, Wester Cleddens Road, Bishopbriggs, Glasgow G64 2NZ, UK

Chapman & Hall GmbH, Pappelallee 3, 69469 Weinheim, Germany

Chapman & Hall USA, 115 Fifth Avenue, New York, NY 10003, USA

Chapman & Hall Japan, ITP-Japan, Kyowa Building, 3F, 2-2-1 Hirakawacho, Chiyoda-ku, Tokyo 102, Japan

Chapman & Hall Australia, 102 Dodds Street, South Melbourne, Victoria 3205, Australia

Chapman & Hall India, R. Seshadri, 32 Second Main Road, CIT East, Madras 600 035, India

First edition 1987, published as *Cyclic Designs*
Second edition 1995

© 1987 J.A. John, 1995 J.A. John and E.R. Williams

Printed in Great Britain by St Edmundsbury Press, Bury St Edmunds, Suffolk

ISBN   0 412 57580 9

A catalogue record for this book is available from the British Library

∞ Printed on permanent acid-free text paper, manufactured in accordance with ANSI/NISO Z39.48-1992 and ANSI/NISO Z39.48-1984 (Permanence of Paper).

Contents

Preface

The biggest single influence on the development of the subject of design of experiments over the past quarter-century has been the availability of computers. Prior to the computer it was essential that any design had a straightforward method of analysis, which meant that the mathematical and combinatorial properties of the designs were of primary importance. Many of the designs proposed and studied also possessed important statistical properties and thus continue to be useful in practice but, now that ease of analysis is less important, a very large number of these designs no longer have any real value. With the advent of the computer it has become possible to study families of designs which have relatively simple methods of construction and which provide large numbers of designs. Within a family, the designs which satisfy certain desirable statistical, rather than mathematical, properties can then be identified using a combination of theory and computing. One of the primary aims of this monograph is to study families of block and row–column designs, for one or more treatment factors, whose methods of construction are cyclical in nature.

Since the first edition of the monograph, previously called *Cyclic Designs*, there have been significant advances in the development of fast computer algorithms to generate efficient block and row–column designs. In particular, the algorithms enable designs to be obtained for more complex blocking structures such as resolvable row–column designs and the recently defined class of Latinized designs. Consequently there is now a greater emphasis on the construction and properties of resolvable block and row–column designs in Chapters 4 and 6 respectively. New results on upper bounds for the average efficiency factors of these designs are also included.

The properties of a design are closely linked to its analysis so that Chapters 1 and 2 develop the theory and criteria needed to judge and compare different designs of the same size. Subsequent chapters are concerned with the construction, properties and analysis of block and row–column designs for one or more treatment factors.

The usual practice adopted in books on the design of experiments is to follow the description of a particular design by its method of analysis and, possibly, a numerical example. While there is still a place for such an approach in introductory texts, the ease and speed with which matrix calculations can now be carried out on computers means that, in practice, separate analyses for different classes of designs are no longer necessary. It is possible for all block designs or all row–column designs, whether for one or more treatment factors, to be analysed using a single computer program. In fact all the designs in this monograph can be analysed using the general approach given in Chapter 7 and which is available in recent versions of the major statistical packages.

The monograph is aimed at the postgraduate level and will hopefully be of interest to students, to those planning and designing experiments and to those engaged in research in the subject. It is assumed that the reader is familiar with the basic principles involved in experimentation, with the requirements for a good experiment, with the concept of factorial experimentation and with some of the standard designs, such as complete block, Latin square and balanced incomplete block designs. Such material is covered in most books on the design of experiments; see, for example, Cox (1958), Fisher (1960) and John and Quenouille (1977). Matrix algebra is used extensively in the monograph. The main results required are given in the Appendix.

We would like to dedicate this monograph to Professor H.D. Patterson of the University of Edinburgh, Scotland. Much of the content is associated with Professor Patterson, either through direct research effort, or collaborative work with the authors. Professor Patterson provided research advice, friendship and constant encouragement to the first author during the writing of the first edition. He was the PhD supervisor for the second author where he again provided continued inspiration and research direction. His influence has therefore largely shaped the overall content of this new edition.

Hamilton and Canberra   J.A. John
1995   E.R. Williams

CHAPTER 1

Block designs

1.1 Introduction

The foundations of the statistical approach to experimentation were laid by R.A. Fisher in the early 1930s. The subject evolved in agriculture but is now applicable to almost all sciences, to engineering and to some arts. The aim of an experiment is to compare a number of treatments on the basis of the responses produced in the experimental material. In agriculture the treatments may be varieties of wheat or different fertilizers, in engineering they may be temperature levels and in a chemical experiment the purpose may be to compare several catalysts. The confidence and accuracy with which treatment differences can be assessed will depend to a large extent on the size of the experiment and on the inherent variability in the experimental material. Dividing the material into relatively homogeneous blocks is an important technique for improving the precision of an experiment.

Suppose that n experimental units or *plots* are available for experimentation and that they are divided into b *blocks* such that the jth block consists of k_j plots $(j = 1, 2, \ldots, b)$. Since blocks are chosen so that the material within them is relatively homogeneous, comparisons made within blocks are usually more accurate than comparisons between blocks. The allocation of treatments to plots should, therefore, be carried out in such a way that important treatment comparisons are made within blocks. The actual allocation of the treatments to plots is called the *design* of the experiment.

A *complete block* design has each of the v treatments occurring once in every block. For such designs all treatment comparisons can be estimated within blocks. For instance, the comparison of two treatments is estimated by the difference between their treatment means, with standard error $\sqrt{(2\sigma^2/r)}$, where σ^2 is the plot variance and r is the number of replications of each treatment. An estimate

Table 1.1. *Analysis of variance for a complete block design*

	Degrees of freedom	Sum of squares	Mean square
Blocks	$b-1$	$(1/k)\sum_{j=1}^{b} B_j^2 - G^2/n$	
Treatments	$v-1$	$(1/r)\sum_{i=1}^{v} T_i^2 - G^2/n$	
Residual	$n-b-v+1$	by subtraction	s^2
Total	$n-1$	$\sum_i \sum_j y_{ij}^2 - G^2/n$	

of σ^2 is given by the residual mean square s^2 in the two-way analysis of variance in Table 1.1, where y_{ij} is the observation on the ith treatment in the jth block, T_i the ith treatment total, B_j the jth block total, and G the overall total of all observations. This analysis of variance can also be used to test for differences between treatment effects; see Section 1.6.

In complete block designs the number of units per block is constant ($k_j = k$, for all j), and equal to the number of treatments, i.e. $k = v$, and consequently $r = b$. However, in many situations restrictions on the experiment, imposed perhaps by the shortage of suitable experimental material or by a large number of treatments under test, may mean that it is not feasible to have a single replication of every treatment in each block. In such cases it becomes necessary to use designs in which the block size is less than the number of treatments, i.e. $k < v$. These designs are called *incomplete block* designs.

In 1936 Yates proposed the use of *balanced incomplete block* and *lattice* designs (Yates, 1936a,b). Balanced incomplete block designs have the property that comparisons between pairs of treatments are all made with the same accuracy. Lattice designs are *resolvable* in the sense that the blocks can be grouped in such a way that each treatment is replicated once in every group. Comparisons between pairs of treatments now vary in accuracy, but fewer replications are required than for many of the balanced designs. Balanced incomplete blocks are discussed in the next section, and lattice designs in Chapter 4.

These designs only exist, however, for a limited number of combinations of v, r and k. When designs of the required size are not available, other incomplete block designs will be required. In Chapters 3 and 4 classes of such designs, primarily based on *cyclical* methods of construction, will be discussed.

In the first instance experiments involving the allocation of a *single set* of treatments to plots will be considered. The treatment comparisons of interest will depend on the composition of this single set and on the purpose of the experiment. For example, a nutritional experiment may be concerned with the effects of a number of diets using litters of mice. The litters constitute blocks since large differences between litters can often be expected. These litter effects are eliminated in any within-litter comparison so that dietary comparisons made within litters are likely to be more accurate than those made between litters. With a completely distinct unstructured set of diets, each diet would probably be compared with every other diet. However, other comparisons may be of interest if there is some structure. If one diet is primarily based on fish while the other diets are free of fish then a comparison of the fish diet with the others may be of special interest.

For a slightly more complex example, consider a subjective aircraft noise experiment where subjects (blocks) are asked to assess the noisiness of four different types of aircraft, namely a supersonic jet, a subsonic jumbo jet and two subsonic smaller jets. Interest may now centre around comparing the supersonic jet with the three subsonic jets and in comparing the jumbo jet with the two smaller jets. For quantitative treatments, a different set of comparisons may be of importance. Suppose a chemical experiment is designed to study the effect of different temperature levels on a certain process. Now it may be of interest to see how yield varied as temperature level increased; whether the change was linear or non-linear. Temperature comparisons which measure linearity or some form of non-linearity would then be examined.

The above are examples of experiments involving single sets of treatments, even though not all treatment comparisons may be of equal interest. More complex treatment structures arise when the experiments involve more than one set of treatments. For example, in a fertilizer experiment the treatments may involve a combination of different amounts of nitrate and potash. The effects of differing amounts of nitrate and potash could be examined separately. In addition, however, the effectiveness of nitrate at different levels of potash could be studied, and vice versa. Such treatment structures give rise to different design problems and will be considered in detail in Chapters 8 and 9.

The analysis of an experiment in which treatment comparisons are estimated within blocks only is called the *intra-block* analysis, and will be considered in this chapter. The allocation of treatments to blocks in some optimal way will be the primary aim of this book. Optimality criteria can be based on the precision with which estimates of treatment comparisons are made in the intra-block analysis; these criteria are discussed in Chapter 2. If the blocks can be regarded as a random sample of blocks from some population then estimates of treatment comparisons may also be available from between block differences; giving rise to an *inter-block* analysis. Where information is available from both between and within blocks the intra- and inter-block estimates can be *combined* to provide overall estimates of treatment comparisons. The problem of how this can be done for block designs, and other blocking structures, is considered in Chapter 7.

Before applying any of the designs obtained by the methods given in this book, they will have to be *randomized*. Randomization ensures the validity of the experiment in the sense that the conclusions are free from the biases of the experimenter. Although randomization is briefly discussed in Section 7.2, a fuller account of the need for randomization and of how different designs are randomized can be found in most of the introductory texts on the design and analysis of experiments. In particular, excellent accounts can be found in Fisher (1960) and Cox (1958).

Finally, it will be convenient to allocate labels to the different treatments. In applying any of the designs, the experimental treatments can then be assigned to these labels. The convention used throughout the book is to label the v treatments $0, 1, \ldots, v-1$.

1.2 Balanced incomplete block designs

Balanced incomplete block designs are designs in which no treatment occurs more than once in any block (i.e. they are *binary*) and in which every pair of treatments occur together in exactly λ (say) blocks. For example a balanced incomplete block design for $v = 5$ treatments, labelled $0, 1, \ldots, 4$, in $b = 5$ blocks with $r = k = 4$ is

$$
\begin{array}{ccccc}
0 & 0 & 0 & 0 & 1 \\
1 & 1 & 1 & 2 & 2 \\
2 & 2 & 3 & 3 & 3 \\
3 & 4 & 4 & 4 & 4
\end{array}
$$

where the blocks are written in columns. Each pair of treatments can be seen to occur together in exactly three blocks, so that $\lambda = 3$.

The parameters v, k, r, b and λ of the balanced incomplete block design satisfy the two relationships

$$bk = vr \tag{1.1}$$

and

$$r(k - 1) = \lambda(v - 1) \tag{1.2}$$

Each side of (1.1) represents the total number of experimental units or plots and is, in fact, a relationship that holds for all block designs. Equation (1.2) is established by noting that a given treatment will occur with $(k - 1)$ other treatments in each of r blocks and also occurs with each of the other $(v - 1)$ treatments in λ blocks.

A balanced incomplete block design cannot exist unless both (1.1) and (1.2) are satisfied. For instance, no design will exist for $v = b = 6$ and $r = k = 3$ since, from (1.2), $\lambda = 6/5$ is not an integer. However, these conditions are not sufficient for the existence of a balanced incomplete block design. Even if both (1.1) and (1.2) are satisfied it does not follow that such a design exists. For example, no balanced incomplete block design exists when $v = 15$, $r = 7$, $k = 5$, $b = 21$ and $\lambda = 2$, even though both conditions are satisfied.

The intra-block analysis of a balanced incomplete block design will be given in Section 1.7.

1.2.1 Construction of balanced incomplete block designs

The balanced incomplete block design for $v = 5$ treatments given above is an example of an *unreduced* design. Such designs are obtained by taking all combinations of the v treatments k at a time. An unreduced design would, therefore, require

$$b = \frac{v!}{k!(v - k)!} \quad \text{blocks and} \quad r = \frac{(v - 1)!}{(k - 1)!(v - k)!} \quad \text{replications}$$

and will have

$$\lambda = \frac{(v - 2)!}{(k - 2)!(v - k)!}$$

Hence, unreduced designs usually require a large number of blocks and replications so that the resulting designs will often be too large for practical purposes. Other balanced incomplete block designs can, however, be constructed which require fewer blocks and replications than the unreduced designs. Two series of designs for $v = m^2$ and $v = m^2 + m + 1$ treatments respectively can be constructed from complete sets of $m \times m$ *mutually orthogonal Latin squares*.

A Latin square is an arrangement of m symbols or labels in an $m \times m$ array such that each label occurs once in each row and once in each column of the array. For example, the following are 4×4 Latin squares, where the labels are denoted by A, B, C and D:

```
A B C D     A B C D     A B C D
B A D C     C D A B     D C B A
C D A B     D C B A     B A D C
D C B A     B A D C     C D A B
```

Two Latin squares are pairwise orthogonal if, when one square is superimposed on the other, each label of one square occurs once with each label of the other square. Three or more squares are mutually orthogonal if they are pairwise orthogonal. The three 4×4 Latin squares above are mutually orthogonal.

For some values of m there will exist a *complete* set of $m - 1$ mutually orthogonal Latin squares. Complete sets are known to exist for any $m = p^s$, where p is a prime number. Tables can be found in Fisher and Yates (1963). The complete sets are used to construct two series of balanced incomplete block designs as follows.

Suppose $v = m^2$ treatments are set out in an $m \times m$ array. A group of m blocks, each of size m, is obtained by letting the rows of the array represent blocks. Another group of m blocks is given by taking the columns of the array as blocks. Now suppose one of the orthogonal squares is superimposed onto the array of treatments. A further group of m blocks will be obtained if all treatments common to a particular label in the square are placed in a block. Each of the $m - 1$ orthogonal squares will produce a set of m blocks in this manner. The resulting design is a balanced incomplete block design with $v = m^2$, $k = m$, $r = m + 1$, $b = m(m + 1)$ and $\lambda = 1$.

Example 1.1

For $m = 3$, a complete set of orthogonal 3×3 squares and the $v = m^2 = 9$ treatments in a 3×3 array are

```
A B C     A B C     0 1 2
C A B     B C A     3 4 5
B C A     C A B     6 7 8
```

Four groups of three blocks are obtained; the first group from the rows, the second group from the columns, and the final two groups from the labels of the two squares respectively. The 12 blocks, again written in columns, are as follows:

```
0  3  6    0  1  2    0  1  2    0  1  2
1  4  7    3  4  5    4  5  3    5  3  4
2  5  8    6  7  8    8  6  7    7  8  6
```

It can be checked that this a balanced incomplete block design with $v = 9$, $b = 12$, $k = 3$, $r = 4$ and $\lambda = 1$.

The *complement* of the design in Example 1.1, obtained by replacing treatments in a block by those not in the block, is also a balanced incomplete block. In general, the complementary design will be a balanced incomplete block design with $v = m^2$, $k = m(m-1)$, $r = m^2 - 1$, $b = m(m+1)$ and $\lambda = m^2 - m + 1$.

The $m(m+1)$ blocks of the designs for $v = m^2$ treatments have been arranged in $m + 1$ groups of m blocks each. Now suppose a new treatment is added to all the blocks in a particular group and that the treatment added is different for each group. Also that one further block is obtained which consists entirely of these $m + 1$ new treatments. This method produces a second series of balanced incomplete block designs with $v = b = m^2 + m + 1$, $r = k = m + 1$ and $\lambda = 1$. Its complement is also a balanced incomplete block design with the same v and b and $r = k = m^2$, $\lambda = m(m-1)$.

Example 1.2

A balanced incomplete block design for $v = b = 13$ and $r = k = 4$, obtained from the design in Example 1.1 for $m = 3$, is given by

```
0   3   6   0    1    2    0    1    2    0    1    2    9
1   4   7   3    4    5    4    5    3    5    3    4    10
2   5   8   6    7    8    8    6    7    7    8    6    11
9   9   9   10   10   10   11   11   11   12   12   12   12
```

Its complement has $r = k = 9$ and $\lambda = 6$.

Unreduced designs and designs obtained using orthogonal squares account for a large number of known balanced incomplete block designs. Various methods have been used to construct other balanced incomplete block designs, and detailed accounts can be found in the books by John (1971) and Raghavarao (1971). Tables of balanced incomplete block designs are given in Fisher and Yates (1963).

1.3 Intra-block model

A single set of v treatments is applied to n experimental plots which have been divided into b blocks with k_j plots in the jth block ($j =$

$1, 2, \ldots, b$). The expected value of the response y_{ijk} obtained when the ith treatment is applied in the jth block on the kth occasion will be assumed to be the sum of three separate components, a general mean parameter μ, a parameter τ_i measuring the effect of the ith treatment and a parameter β_j giving the effect of the jth block. It will be assumed that the responses are uncorrelated, with the same variance σ^2.

If n_{ij} is the number of times that the ith treatment is applied to the jth block, then the model can be written as

$$y_{ijk} = \mu + \tau_i + \beta_j + \varepsilon_{ijk} \qquad (1.3)$$
$$(i = 1, 2, \ldots, v; j = 1, 2, \ldots, b; k = 1, 2, \ldots, n_{ij})$$

where the error terms ε_{ijk} are uncorrelated random variables each with mean zero and variance σ^2. In the intra-block analysis the block effects are taken to be a fixed set of parameters, and are not to be regarded as random variables. One of the primary objectives of the analysis will be to obtain estimates of the treatment parameters τ_i, together with the standard errors of these estimates. It may also be important to establish whether the treatment comparisons of interest differ significantly from zero. The additive model (1.3) will be assumed to be an appropriate model for block designs. In practice, of course, it is important that the assumptions underlying this model are carefully examined. Much useful information can be obtained by studying the differences between the observed responses and those obtained from the fitted model, i.e. from an analysis of the residuals. Details of such analyses can be found in many statistical texts, e.g. Draper and Smith (1981) and John and Quenouille (1977).

In matrix notation, model (1.3) can be written as

$$\mathbf{y} = \mathbf{W}\boldsymbol{\alpha} + \boldsymbol{\varepsilon} \qquad (1.4)$$

where $\mathbf{y}$ and $\boldsymbol{\varepsilon}$ are $n \times 1$ vectors of response and error terms respectively, $\boldsymbol{\alpha}$ is a $p \times 1$ vector of parameters and $\mathbf{W}$ an $n \times p$ design matrix of ones and zeros, and where $p = 1 + v + b$. It is assumed that $\mathrm{E}(\boldsymbol{\varepsilon})$, the expected value of the error vector, is zero and that $\mathrm{V}(\boldsymbol{\varepsilon})$, the variance–covariance matrix of the error vector, is $\sigma^2 \mathbf{I}$.

Many of the features of this model and of the subsequent algebra can be illustrated by following through the various stages with a hypothetical example of a small block design. The example is used for illustrative purposes only and it is not necessarily suggested that such a design be used in practice.

Table 1.2. *Responses for the hypothetical experiment*

Treatment	Block			
	1	2	3	4
A	8		13	9
B		9	16	8
C	14	5	12	

Example 1.3

An experiment is carried out with three treatments set out in four blocks. The responses obtained are shown in Table 1.2.

Letting τ_1, τ_2 and τ_3 be the treatment parameters corresponding to treatments A, B and C respectively, the model assumed for this experiment is, using the matrix form given in (1.4), as follows:

$$
\begin{pmatrix} 8 \\ 14 \\ 9 \\ 5 \\ 13 \\ 16 \\ 12 \\ 9 \\ 8 \end{pmatrix} = \begin{pmatrix} 1 & | & 1\ 0\ 0 & | & 1\ 0\ 0\ 0 \\ 1 & | & 0\ 0\ 1 & | & 1\ 0\ 0\ 0 \\ 1 & | & 0\ 1\ 0 & | & 0\ 1\ 0\ 0 \\ 1 & | & 0\ 0\ 1 & | & 0\ 1\ 0\ 0 \\ 1 & | & 1\ 0\ 0 & | & 0\ 0\ 1\ 0 \\ 1 & | & 0\ 1\ 0 & | & 0\ 0\ 1\ 0 \\ 1 & | & 0\ 0\ 1 & | & 0\ 0\ 1\ 0 \\ 1 & | & 1\ 0\ 0 & | & 0\ 0\ 0\ 1 \\ 1 & | & 0\ 1\ 0 & | & 0\ 0\ 0\ 1 \end{pmatrix} \begin{pmatrix} \mu \\ \tau_1 \\ \tau_2 \\ \tau_3 \\ \beta_1 \\ \beta_2 \\ \beta_3 \\ \beta_4 \end{pmatrix} + \begin{pmatrix} \varepsilon_1 \\ \varepsilon_2 \\ \varepsilon_3 \\ \varepsilon_4 \\ \varepsilon_5 \\ \varepsilon_6 \\ \varepsilon_7 \\ \varepsilon_8 \\ \varepsilon_9 \end{pmatrix}
$$

Note that the columns of the **W** matrix can be conveniently partitioned as indicated into three submatrices corresponding to the parameter μ, the treatment parameters τ_1, τ_2 and τ_3 and the block parameters β_1, β_2, β_3 and β_4.

Hence, the matrix **W** and the vector $\boldsymbol{\alpha}$ in (1.4) can be partitioned as follows:

$$
\mathbf{W} = (\mathbf{1} \,|\, \mathbf{X} \,|\, \mathbf{Z})
$$
$$
\boldsymbol{\alpha} = (\mu \,|\, \boldsymbol{\tau}' \,|\, \boldsymbol{\beta}')'
$$

where **1** is a column vector of ones and where **X** and **Z** are the $n \times v$ and $n \times b$ *design* matrices for treatments and blocks respectively. The model (1.4) can now be rewritten as

$$
\mathbf{y} = \mathbf{1}\mu + \mathbf{X}\boldsymbol{\tau} + \mathbf{Z}\boldsymbol{\beta} + \boldsymbol{\varepsilon} \tag{1.5}
$$

Some further notation will be introduced to simplify the algebra of subsequent sections. In the treatment design matrix **X** there is a single column for each treatment. The sum (and sum of squares) of the elements in any column gives the number of times the corresponding treatment occurs in the design. Further, the sum of cross products of any two columns must be zero since any plot can only contain one treatment at a time. Hence, if $\mathbf{r}$ is the *replication vector*, whose elements give the number of times each treatment occurs in the design, then

$$\mathbf{X'1} = \mathbf{r} \quad \text{and} \quad \mathbf{X'X} = \mathbf{r}^\delta \tag{1.6}$$

where $\mathbf{r}^\delta$ is the diagonal matrix whose diagonal elements are those of the vector $\mathbf{r}$.

Similarly, if $\mathbf{k}$ is a vector whose jth element is k_j, the number of plots in the jth block, then

$$\mathbf{Z'1} = \mathbf{k} \quad \text{and} \quad \mathbf{Z'Z} = \mathbf{k}^\delta \tag{1.7}$$

It is readily verified in Example 1.3 that these calculations give $\mathbf{r} = (3\ 3\ 3)'$ and $\mathbf{k} = (2\ 2\ 3\ 2)'$.

The sum of cross products of the ith column of $\mathbf{X}$ and the jth column of $\mathbf{Z}$ will give the number of times the ith treatment occurs in the jth block, namely n_{ij}. If $\mathbf{N}$ is the matrix whose (ij)th element is n_{ij}, then

$$\mathbf{X'Z} = \mathbf{N} \tag{1.8}$$

The $v \times b$ matrix $\mathbf{N}$, which is called the *incidence matrix* of the design, has a row for each treatment and a column for each block. For Example 1.3,

$$\mathbf{N} = \begin{pmatrix} 1 & 0 & 1 & 1 \\ 0 & 1 & 1 & 1 \\ 1 & 1 & 1 & 0 \end{pmatrix}$$

Finally, if $\mathbf{T}$ and $\mathbf{B}$ are vectors of treatment and block totals respectively and if G is the overall total, then from a vector multiplication of the columns of $\mathbf{W}$ with the response vector $\mathbf{y}$ it is established that

$$\mathbf{X'y} = \mathbf{T}, \qquad \mathbf{Z'y} = \mathbf{B}, \qquad \mathbf{1'y} = G \tag{1.9}$$

Again for Example 1.3, $\mathbf{T} = (30\ 33\ 31)'$, $\mathbf{B} = (22\ 14\ 41\ 17)'$ and $G = 94$.

1.4 Least squares analysis

The least squares estimator of the parameter vector $\boldsymbol{\alpha}$ in (1.4) is obtained by minimizing the error sum of squares $\boldsymbol{\varepsilon}'\boldsymbol{\varepsilon}$ with respect to α. Since

$$\varepsilon'\varepsilon = (\mathbf{y} - \mathbf{W}\alpha)'(\mathbf{y} - \mathbf{W}\alpha) = \mathbf{y}'\mathbf{y} - 2\mathbf{y}'\mathbf{W}\alpha + \alpha'\mathbf{W}'\mathbf{W}\alpha$$

then

$$\partial(\varepsilon'\varepsilon)/\partial\alpha = -2\mathbf{W}'\mathbf{y} + 2\mathbf{W}'\mathbf{W}\alpha$$

Equating these partial derivatives to zero gives the *normal equations* $\mathbf{W}'\mathbf{W}\hat{\alpha} = \mathbf{W}'\mathbf{y}$ with a solution of these equations producing an estimate $\hat{\alpha}$ of the true parameter vector α. Partitioning $\mathbf{W}$ and α as in the previous section gives the normal equations, for the model (1.5), as

$$n\hat{\mu} + \mathbf{r}'\hat{\tau} + \mathbf{k}'\hat{\beta} = G \qquad (1.10)$$

$$\mathbf{r}\hat{\mu} + \mathbf{r}^{\delta}\hat{\tau} + \mathbf{N}\hat{\beta} = \mathbf{T} \qquad (1.11)$$

$$\mathbf{k}\hat{\mu} + \mathbf{N}'\hat{\tau} + \mathbf{k}^{\delta}\hat{\beta} = \mathbf{B} \qquad (1.12)$$

where use has been made of (1.6)–(1.9).

These equations do not have a unique solution since they are less than full rank. This can be seen by noting that the sum of the v rows of (1.11) and the sum of the b rows of (1.12) both give equation (1.10). Thus, the rank of the matrix of coefficients $\mathbf{W}'\mathbf{W}$ is at least two less than the size of the matrix, i.e. rank($\mathbf{W}'\mathbf{W}$) $\leq v + b - 1$.

A solution, or solutions, to the normal equations can, however, be readily found. One method, for instance, is to put a set of parameters equal to zero, or to other arbitrary values, in such a way that the remaining equations are of full rank. In general, any solution to the normal equations $\mathbf{W}'\mathbf{W}\hat{\alpha} = \mathbf{W}'\mathbf{y}$ can be written as $\hat{\alpha} = (\mathbf{W}'\mathbf{W})^{-}\mathbf{W}'\mathbf{y}$ where $(\mathbf{W}'\mathbf{W})^{-}$ is a generalized inverse of $\mathbf{W}'\mathbf{W}$, satisfying

$$\mathbf{W}'\mathbf{W}(\mathbf{W}'\mathbf{W})^{-}\mathbf{W}'\mathbf{W} = \mathbf{W}'\mathbf{W}.$$

A brief discussion of generalized inverses is given in the Appendix.

The task of finding a solution in the present case is simplified if a set of equations involving only the treatment parameters (the parameters of interest) is obtained by eliminating $\hat{\mu}$ and $\hat{\beta}$ from the normal equations. This is achieved by first premultiplying (1.12) by $\mathbf{Nk}^{-\delta}$, where $\mathbf{k}^{-\delta}$ is the inverse of the matrix $\mathbf{k}^{\delta}$, to give

$$\mathbf{r}\hat{\mu} + \mathbf{Nk}^{-\delta}\mathbf{N}'\hat{\tau} + \mathbf{N}\hat{\beta} = \mathbf{Nk}^{-\delta}\mathbf{B}$$

and then subtracting this equation from (1.11) to give

$$\mathbf{A}\hat{\tau} = \mathbf{q} \qquad (1.13)$$

where

$$A = r^\delta - Nk^{-\delta}N' \tag{1.14}$$

and

$$q = T - Nk^{-\delta}B \tag{1.15}$$

A solution to the normal equations can now be obtained by solving the equations (1.13) for $\hat{\tau}$ and then from (1.12) obtaining $\hat{\beta}$ as

$$\hat{\beta} = k^{-\delta}B - k^{-\delta}N'\hat{\tau} - 1\hat{\mu} \tag{1.16}$$

An intuitively appealing estimate of μ is given by $\hat{\mu} = \bar{y}$, although any other value can be used.

Hence the problem of finding a solution to the normal equations has now been simplified to one of finding a solution to (1.13).

The equations (1.13) are called the *reduced normal equations* for the treatment parameters obtained by eliminating the mean and block parameters from the full set of normal equations. The $v \times v$ coefficient matrix A is called the *information matrix* of the design in the intra-block analysis and plays a key role, in particular in establishing suitable optimality criteria for choosing between different block designs, as will be shown in the next chapter. The vector q is called the vector of *adjusted treatment totals* since linear combinations of the block totals B are subtracted from the treatment totals T. Note that the vector involves linear combinations of the responses since it can be written as

$$q = (X' - Nk^{-\delta}Z')y = X'\{I - Z(Z'Z)^{-1}Z'\}y$$

Again no unique solution of the reduced normal equations exists since the information matrix A is not of full rank. The rows and columns of A sum of zero, so that $\text{rank}(A) \leq v - 1$. All solutions to the equations (1.13) can be written as

$$\hat{\tau} = \Omega q \tag{1.17}$$

where Ω is a generalized inverse of A satisfying $A\Omega A = A$.

A generalized inverse is readily obtained if A is expressed in canonical form. Let the columns of A be spanned by a set of normalized eigenvectors $p_1, p_2, \ldots, p_v$ with corresponding eigenvalues $\theta_1, \theta_2, \ldots, \theta_v$; $p_i'p_i = 1$ and $p_i'p_j = 0$ $(i \neq j)$. Then, in canonical form

$$A = \sum_{i=1}^{v} \theta_i p_i p_i' \tag{1.18}$$

One choice for Ω is given by

$$A^+ = \sum \theta_i^{-1} p_i p_i' \tag{1.19}$$

where the summation is over all the non-zero eigenvalues. A^+ is the unique Moore–Penrose generalized inverse of A; see the Appendix.

Since $\hat{\tau}'1 = 0$ if $\hat{\tau}$ is obtained from (1.17) with Ω given by A^+ in (1.19), it is suggested that $\hat{\mu} = \bar{y}$ is added to each element in $\hat{\tau}$ so that the treatment estimates from the intra-block analysis are then directly comparable with the unadjusted treatment means $r^{-\delta}T$. Hence, the estimates are

$$\hat{\tau}^* = \hat{\tau} + \bar{y}1 \tag{1.20}$$

Example 1.3 (continued)

$$Nk^{-\delta}N' = \frac{1}{6}\begin{pmatrix} 8 & 5 & 5 \\ 5 & 8 & 5 \\ 5 & 5 & 8 \end{pmatrix}, \quad Nk^{-\delta}B = \frac{1}{6}\begin{pmatrix} 199 \\ 175 \\ 190 \end{pmatrix}$$

Hence

$$A = r^\delta - Nk^{-\delta}N' = \frac{5}{6}\begin{pmatrix} 2 & -1 & -1 \\ -1 & 2 & -1 \\ -1 & -1 & 2 \end{pmatrix}$$

$$q = T - Nk^{-\delta}B = \frac{1}{6}\begin{pmatrix} -19 \\ 23 \\ -4 \end{pmatrix}$$

Normalized eigenvectors of A are given by

$$p_1 = \frac{1}{\sqrt{(2)}}\begin{pmatrix} 1 \\ 0 \\ -1 \end{pmatrix}, \quad p_2 = \frac{1}{\sqrt{(6)}}\begin{pmatrix} 1 \\ -2 \\ 1 \end{pmatrix}, \quad p_3 = \frac{1}{\sqrt{(3)}}\begin{pmatrix} 1 \\ 1 \\ 1 \end{pmatrix}$$

with eigenvalues $\theta_1 = 15/6$, $\theta_2 = 15/6$, $\theta_3 = 0$. Therefore

$$A^+ = \frac{6}{15}\left\{ \frac{1}{2}\begin{pmatrix} 1 \\ 0 \\ -1 \end{pmatrix}(1 \; 0 \; -1) + \frac{1}{6}\begin{pmatrix} 1 \\ -2 \\ 1 \end{pmatrix}(1 \; -2 \; 1) \right\}$$

$$= \frac{2}{15}\begin{pmatrix} 2 & -1 & -1 \\ -1 & 2 & -1 \\ -1 & -1 & 2 \end{pmatrix}$$

and

$$\hat{\tau} = A^+q = \frac{1}{15}\begin{pmatrix} -19 \\ 23 \\ -4 \end{pmatrix}$$

Finally

$$\hat{\mu} = \bar{y} = \frac{94}{9}$$

and

$$\hat{\beta} = \mathbf{k}^{-\delta}(\mathbf{B} - \mathbf{N}'\hat{\tau}) - \mathbf{1}\bar{y} = \frac{1}{90} \begin{pmatrix} 119 \\ -367 \\ 290 \\ -187 \end{pmatrix}$$

1.5 Estimability

Properties of the parameter estimators obtained from a solution of the normal equations will now be examined. First, the expected value and variance–covariance matrix of the adjusted treatment vector $\mathbf{q}$ will be obtained. Let the model (1.5) be premultiplied successively by $\mathbf{X}'$ and $\mathbf{N}\mathbf{k}^{-\delta}\mathbf{Z}'$ to give

$$\mathbf{X}'\mathbf{y} = \mathbf{r}\mu + \mathbf{r}^{\delta}\tau + \mathbf{N}\beta + \mathbf{X}'\varepsilon \qquad (1.21)$$
$$\mathbf{N}\mathbf{k}^{-\delta}\mathbf{Z}'\mathbf{y} = \mathbf{r}\mu + \mathbf{N}\mathbf{k}^{-\delta}\mathbf{N}'\tau + \mathbf{N}\beta + \mathbf{N}\mathbf{k}^{-\delta}\mathbf{Z}'\varepsilon \quad (1.22)$$

where again use has been made of (1.6)–(1.8). Subtracting (1.22) from (1.21) gives

$$\mathbf{q} = \mathbf{A}\tau + (\mathbf{X}' - \mathbf{N}\mathbf{k}^{-\delta}\mathbf{Z}')\varepsilon \qquad (1.23)$$

Since $E(\varepsilon) = \mathbf{0}$ and $V(\varepsilon) = \sigma^2\mathbf{I}$ it follows that

$$E(\mathbf{q}) = \mathbf{A}\tau$$

and

$$\begin{aligned} V(\mathbf{q}) &= (\mathbf{X}' - \mathbf{N}\mathbf{k}^{-\delta}\mathbf{Z}')(\mathbf{X}' - \mathbf{N}\mathbf{k}^{-\delta}\mathbf{Z}')'\sigma^2 \\ &= (\mathbf{r}^{\delta} - \mathbf{N}\mathbf{k}^{-\delta}\mathbf{N}')\sigma^2 = \mathbf{A}\sigma^2 \end{aligned}$$

so that, using (1.17)

$$E(\hat{\tau}) = \mathbf{\Omega}\mathbf{A}\tau \qquad (1.24)$$
$$V(\hat{\tau}) = \mathbf{\Omega}\mathbf{A}\mathbf{\Omega}\sigma^2 \qquad (1.25)$$

Hence $\hat{\tau}$ is an unbiased estimator of a function of the treatment parameters. Further, this function depends on the non-unique generalized inverse matrix $\mathbf{\Omega}$. Although the treatment parameters themselves are not uniquely estimated and do not have unbiased estimators, certain linear functions of these treatment parameters are *estimable*. In general, a linear function $\mathbf{c}'\tau$ is said to be estimable if there exists some linear combination of the responses $\mathbf{b}'\mathbf{y}$ such that $E(\mathbf{b}'\mathbf{y}) = \mathbf{c}'\tau$. Since $\hat{\tau}$ is a linear function of the responses it follows from (1.24) that $c'\tau$ is estimable if the coefficient vector c is chosen to satisfy $c' = c'\Omega A$. This *estimability condition* is, in fact, both necessary and sufficient for $c'\tau$ to be estimable; for a fuller discussion of estimability see Searle (1971, p. 180). An unbiased estimator of the estimable function $c'\tau$ is $c'\hat{\tau}$ and, further, this estimator is invariant to whatever solution of the reduced normal equations (1.13) is used for $\hat{\tau}$. Finally, from (1.25) it follows that the variance of this estimator is given by

$$\text{var}(c'\hat{\tau}) = c'\Omega c \sigma^2 \tag{1.26}$$

As an example of an estimable function, consider again the canonical forms of A and Ω given in (1.18) and (1.19) respectively. Let p_1 be an eigenvector with non-zero eigenvalue θ_1. It follows immediately that $p_1' = p_1'\Omega A$ so that $p_1'\tau$ is estimable with unbiased estimator $\theta_1^{-1}p_1'q$, whose variance is $\theta_1^{-1}\sigma^2$.

1.6 Analysis of variance

The further assumption that the error terms ε in (1.5) are normally distributed is required when carrying out significance tests or setting confidence intervals. It will be assumed, therefore, that the error vector ε follows a multivariate normal distribution with mean vector 0 and variance–covariance matrix $\sigma^2 I$.

The total sum of squares $y'y$, for the general linear model $y = W\alpha + \varepsilon$, can be partitioned into two components as follows:

$$y'y = \hat{\alpha}'W'y + (y - W\hat{\alpha})'(y - W\hat{\alpha}) \tag{1.27}$$

where $\hat{\alpha}$ is any solution to the normal equations $W'W\alpha = W'y$. Since $y - W\hat{\alpha}$ is the vector of residuals, the second component in (1.27) is the residual sum of squares. The first component, therefore, represents that part of the total variation accounted for by the model. That is, the sum of squares due to fitting all the parameters in the model $y = W\alpha + \varepsilon$ is given by $\hat{\alpha}'W'y = S(\alpha)$. Note that this sum of squares is the sum of cross products of the least squares estimator $\hat{\alpha}$ and the terms on the right-hand side of the normal equations $W'y$.

Hence, the sum of squares due to fitting the parameters in the intra-block model (1.5) is given by

$$S(\mu, \tau, \beta) = \hat{\mu}G + \hat{\tau}'T + \hat{\beta}'B$$

which, using (1.15) and (1.16), becomes

$$S(\mu, \tau, \beta) = \hat{\tau}'q + B'k^{-\delta}B \tag{1.28}$$

Suppose now that the hypothesis of no significant differences between treatment parameters is to be tested. That is, the null hypothesis of interest is

$$H_0 : \tau_1 = \tau_2 = \ldots = \tau_v$$

Under the assumption that H_0 is true, model (1.5) reduces to the one-way model given by

$$\mathbf{y} = \mathbf{1}\mu + \mathbf{Z}\beta + \varepsilon \qquad (1.29)$$

The normal equations for this model are

$$\begin{aligned} n\hat{\mu} + \mathbf{k}'\hat{\beta} &= G \\ \mathbf{k}\hat{\mu} + \mathbf{k}^{\delta}\hat{\beta} &= \mathbf{B} \end{aligned}$$

so that with $\hat{\mu} = \bar{y}$,

$$\hat{\beta} = \mathbf{k}^{-\delta}\mathbf{B} - \mathbf{1}\bar{y}$$

Thus

$$S(\mu, \beta) = \hat{\mu}G + \hat{\beta}'\mathbf{B} = \mathbf{B}'\mathbf{k}^{-\delta}\mathbf{B} \qquad (1.30)$$

The difference between the two sums of squares (1.28) and (1.30) is the sum of squares due to testing the hypothesis H_0. Since it is the sum of squares due to treatment effects after eliminating the overall mean parameter and the block parameters it will be denoted by $S(\tau/\mu, \beta)$. Hence

$$S(\tau/\mu, \beta) = S(\mu, \tau, \beta) - S(\mu, \beta) = \hat{\tau}'\mathbf{q} \qquad (1.31)$$

The sum of squares $S(\mu, \beta)$ in (1.31) can itself be partitioned further. Putting all block parameters equal in (1.29) leads to the reduced model $\mathbf{y} = \mathbf{1}\mu + \varepsilon$. From the normal equations for this model it follows that $\hat{\mu} = \bar{y}$ and $S(\mu) = G^2/n$. Hence

$$S(\beta/\mu) = S(\mu, \beta) - S(\mu) = \mathbf{B}'\mathbf{k}^{-\delta}\mathbf{B} - G^2/n \qquad (1.32)$$

$S(\beta/\mu)$ is the sum of squares due to blocks after eliminating the mean. Since it does not refer at all to treatments it represents between block variation; it can be seen in (1.32) to be a function of the block totals $\mathbf{B}$. $S(\tau/\mu, \beta)$, on the other hand, represents variation within blocks since block effects have been eliminated.

The above results are summarized in the *analysis of variance* given in Table 1.3. It should be noted that all treatment differences must be estimable in the design if H_0 is to be tested; this point will be discussed in Section 1.8. The residual mean square s^2 provides an unbiased estimator of σ^2 and is used as the estimate of experimental error in any significance test and in setting confidence

Table 1.3. *Intra-block analysis of variance*

Source of variation	Degrees of freedom (d.f.)	Sum of squares (s.s.)	Mean square (m.s.)
Between blocks (ignoring treatments)	$b-1$	$S(\beta/\mu) = \mathbf{B}'\mathbf{k}^{-\delta}\mathbf{B}$ $- G^2/n$	
Between treatments (eliminating blocks)	$v-1$	$S(\tau/\mu,\beta) = \hat{\tau}'\mathbf{q}$	$\frac{\hat{\tau}'\mathbf{q}}{v-1}$
Residual	$n-b-v+1$	$\mathbf{y}'\mathbf{y} - S(\mu,\tau,\beta)$	s^2
Total	$n-1$	$\mathbf{y}'\mathbf{y} - G^2/n$	

limits. The hypothesis H_0 is then tested by comparing the variance ratio

$$F = \frac{\hat{\tau}'\mathbf{q}/(v-1)}{s^2}$$

with the appropriate percentage points of the F distribution with $v-1$ and $n-b-v+1$ degrees of freedom.

To produce the analysis of variance given in Table 1.3 a sequence of models was fitted, starting with a mean term then adding in turn the block parameters and the treatment parameters. This led to the following decomposition of the sum of squares $S(\mu,\tau,\beta)$ due to fitting the intra-block model (1.5):

$$S(\mu,\tau,\beta) = S(\mu) + S(\beta/\mu) + S(\tau/\mu,\beta) \qquad (1.33)$$

This particular model fitting sequence was followed as interest is centred on the treatments in a blocking experiment and, thus, it is necessary to estimate and test the treatment parameters in the absence of, or after adjusting for, the block effects.

However, adding treatment parameters to the model before the block parameters would lead to another analysis of variance and the following decomposition:

$$S(\mu,\tau,\beta) = S(\mu) + S(\tau/\mu) + S(\beta/\mu,\tau) \qquad (1.34)$$

Analogous to (1.32),

$$S(\tau/\mu) = \mathbf{T}'\mathbf{r}^{-\delta}\mathbf{T} - G^2/n \qquad (1.35)$$

and represents the unadjusted treatment sum of squares (between treatments, ignoring blocks). By subtraction from (1.33) and (1.34) the adjusted block sum of squares (between blocks, eliminating treatments) is given by

$$S(\beta/\mu, \tau) = S(\beta/\mu) + S(\tau/\mu, \beta) - S(\tau/\mu) \qquad (1.36)$$

Alternatively, following through similar algebra to that used to obtain (1.31), it can be shown that

$$S(\beta/\mu, \tau) = \hat{\beta}' \mathbf{w} \qquad (1.37)$$

where $\mathbf{w}$ is the vector of adjusted block totals given by

$$\mathbf{w} = \mathbf{B} - \mathbf{N}' \mathbf{r}^{-\delta} \mathbf{T}$$

and $\hat{\beta}$ is a solution to the set of equations

$$(\mathbf{k}^\delta - \mathbf{N}' \mathbf{r}^{-\delta} \mathbf{N})\hat{\beta} = \mathbf{w}$$

In a blocking experiment the second decomposition given in (1.34) is of little importance in itself since it is rarely of interest to examine block effects in detail. However, the sums of squares given in (1.35), (1.36) and (1.37) will be needed in the discussions of orthogonality in Section 1.9 and of the recovery of inter-block information in Chapter 7; it is convenient to define them here.

Example 1.3 (continued)
Using the data and calculations of Sections 1.3 and 1.4,

$$
\begin{aligned}
\mathbf{y}'\mathbf{y} &= 8^2 + 14^2 + 9^2 + 5^2 + 13^2 + 16^2 + 12^2 + 9^2 + 8^2 \\
&= 1080 \\
S(\mu) &= G^2/n = 94^2/9 = 981.78 \\
S(\beta/\mu) &= \frac{22^2}{2} + \frac{14^2}{2} + \frac{41^2}{3} + \frac{17^2}{2} - \frac{G^2}{n} = 63.06 \\
S(\tau/\mu, \beta) &= \frac{1}{15 \times 6} \left(\begin{array}{ccc} -19 & 23 & -4 \end{array} \right) \left(\begin{array}{c} -19 \\ 23 \\ -4 \end{array} \right) = 10.07
\end{aligned}
$$

The intra-block analysis of variance is shown in Table 1.4. The variance ratio statistic to test $H_0 : \tau_1 = \tau_2 = \tau_3$ is

$$F = \frac{5.04}{8.36} = 0.6$$

based on two and three degrees of freedom. As this is a small hypothetical example with very few degrees of freedom for error, no conclusions should be drawn from this analysis. The above arithmetic is purely to illustrate the calculations involved.

Table 1.4. *Analysis of variance table*

	d.f.	s.s.	m.s.
Between blocks (ignoring treatments)	3	63.06	21.02
Between treatments (eliminating blocks)	2	10.07	5.04
Residual	3	25.09	8.36
Total	8	98.22	

Finally, the unadjusted treatment sum of squares and the adjusted block sum of squares are given by

$$S(\tau/\mu) = \frac{30^2 + 33^2 + 31^2}{3} - \frac{G^2}{n} = 1.56$$
$$S(\beta/\mu, \tau) = 63.06 + 10.07 - 1.56 = 71.57$$

Hypotheses other than H_0 involving the treatment parameters may be of interest. It may be desired to test the hypothesis that some estimable function $c'\tau$ is equal to some given quantity. More generally, the hypothesis may involve a number of such estimable functions. Suppose that C is a matrix such that each of the rows of $C\tau$ is estimable and linearly independent of the other rows. Then the sum of squares due to testing the hypothesis $H_0(C) : C\tau = m$, where m is a vector of known quantities, is given by

$$S\{H_0(C)\} = (C\hat{\tau} - m)'(C\Omega C')^{-1}(C\hat{\tau} - m) \qquad (1.38)$$

based on ν degrees of freedom, where $\nu = \text{rank}(C)$; see Searle (1971, p. 190). In Chapters 8 and 9 on factorial experiments it will be more convenient to work with C matrices which do not have full row rank. In such cases, the sum of squares is again given by (1.38) with the inverse of $C\Omega C'$ now replaced by a generalized inverse of $C\Omega C'$; see John and Smith (1974).

1.7 Intra-block analysis of a balanced incomplete block design

A balanced incomplete block design has v treatments set out in b blocks each of size k with each treatment replicated r times. Also the diagonal elements of the concurrence matrix NN' are all equal to r and the off-diagonal elements are all equal to λ, i.e.

$$\mathbf{N}\mathbf{N}' = (r - \lambda)\mathbf{I} + \lambda\mathbf{J}$$

The information matrix $\mathbf{A}$ of (1.14) is then given by

$$\mathbf{A} = (\lambda v/k)(\mathbf{I} - \mathbf{K}) \tag{1.39}$$

where $\mathbf{K} = (1/v)\mathbf{J}$ and $\mathbf{J}$ is a $v \times v$ matrix with every element equal to 1.

Since any contrast vector $\mathbf{c}$ is an eigenvector of $\mathbf{A}$ with eigenvalue $\lambda v/k$, it follows that the Moore–Penrose generalized inverse (1.19) is given by

$$\mathbf{A}^{+} = (k/\lambda v)(\mathbf{I} - \mathbf{K}) \tag{1.40}$$

Hence, $\mathbf{c}'\boldsymbol{\tau}$ is estimated by

$$\mathbf{c}'\hat{\boldsymbol{\tau}} = (k/\lambda v)\mathbf{c}'\mathbf{q} \tag{1.41}$$

with

$$\mathrm{var}(\mathbf{c}'\hat{\boldsymbol{\tau}}) = (k/\lambda v)\mathbf{c}'\mathbf{c}\sigma^2 \tag{1.42}$$

Thus comparisons between pairs of treatments are all estimated with the same accuracy.

Further, the $v - 1$ degrees of freedom for the treatment sum of squares, adjusted for blocks, can be partitioned into $v - 1$ components each with a single degree of freedom by using any set of $v - 1$ orthogonal contrasts. Thus, the adjusted treatment sum of squares (1.31) can be written as

$$\hat{\boldsymbol{\tau}}'\mathbf{q} = (k/\lambda v)\mathbf{q}'\mathbf{q} = \sum_{i=1}^{v-1}(k/\lambda v)(\mathbf{c}_i'\mathbf{q})^2/(\mathbf{c}_i'\mathbf{c}_i) \tag{1.43}$$

for any set of contrast vectors $\mathbf{c}_i$ satisfying $\mathbf{c}_i'\mathbf{c}_j = 0$ $(i, j = 1, 2, \ldots, v - 1; i \neq j)$.

In Section 1.2.1 it was stated that the complement of a particular class of balanced incomplete block designs are also balanced incomplete block designs. This result can easily be shown to be true for any balanced incomplete block design. Let $\mathbf{N}$ and $\mathbf{N}_c$ be the incidence matrices of the design and its complement respectively. Then, by definition, $\mathbf{N}_c = \mathbf{J} - \mathbf{N}$. Hence

$$\mathbf{N}_c\mathbf{N}_c' = \mathbf{N}\mathbf{N}' + (b - 2r)\mathbf{J}$$

which is the concurrence matrix of a balanced incomplete block design in which each pair of treatments occurs together in $(\lambda + b - 2r)$ blocks.

1.8 Connectedness

The $v \times v$ information matrix $\mathbf{A}$ given in (1.14) has rank less than or equal to $v - 1$. The equality will hold if and only if the design is *connected*. In the discussion which follows attention is initially restricted to *binary* designs which have equal replication and equal block sizes, i.e. $r_i = r$ and $k_j = k$ for all i and j. A binary design is one in which each element of the incidence matrix $\mathbf{N}$ is either 0 or 1. For such designs the information matrix simplifies to

$$\mathbf{A} = r\mathbf{I} - (1/k)\mathbf{N}\mathbf{N}' \qquad (1.44)$$

where the treatment *concurrence matrix* $\mathbf{N}\mathbf{N}'$ has diagonal elements equal to r and off-diagonal elements equal to the number of times pairs of treatments occur together in a block. Hence, if $\mathbf{A} = (a_{ij})$ then

$$a_{ij} = \begin{cases} r(k-1)/k, & i = j \\ -\lambda_{ij}/k, & i \neq j \end{cases}$$

where λ_{ij} gives the number of blocks containing both the ith and jth treatments. Note that for balanced incomplete block designs, $\lambda_{ij} = \lambda$ for all i and j.

To introduce the concept of connectedness consider the following example:

Example 1.4
Two designs for six treatments in six blocks of size two ($k = 2$) with each treatment replicated twice ($r = 2$) are

$$\begin{array}{ccccccc} \text{Design 1}: & 0 & 1 & 2 & 3 & 4 & 5 \\ & 2 & 3 & 4 & 5 & 0 & 1 \end{array}$$

$$\begin{array}{ccccccc} \text{Design 2}: & 0 & 1 & 2 & 3 & 4 & 5 \\ & 1 & 2 & 3 & 4 & 5 & 0 \end{array}$$

The six blocks in Design 1 can be split into two groups of three blocks each, with treatments 0, 2 and 4 in the blocks of one group and treatments 1, 3 and 5 in the other group. There is no treatment common to the two groups. Design 1 is an example of a disconnected design. The blocks in Design 2 cannot be split into distinct groups of blocks in this manner. With this design all treatments can be compared within blocks, either directly or indirectly via other blocks. Design 2 is a connected design.

Further insight into the concept of connectedness can be obtained if the designs are represented graphically. Let the points

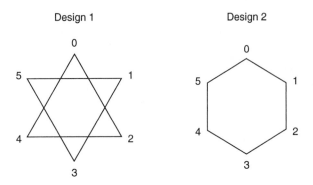

Figure 1.1. *Graphs for the designs in Example 1.4*

(vertices) of the graph represent treatments and let a line (edge) join two points if the corresponding treatments occur together in a block. Such a graph is called the *treatment concurrence graph* of the design since the number of lines joining any two points are given by the corresponding elements of the treatment concurrence matrix $\mathbf{NN'}$. The graphs for the two designs are shown in Fig. 1.1. In the connected design (Design 2) it is possible to find a path from any point in the treatment concurrence graph to any other point. In the disconnected design (Design 1) this is not possible, since no path joins any even to any odd numbered point.

It will now be shown that a design is connected if and only if rank$(\mathbf{A}) = v - 1$. Suppose that rank$(\mathbf{A}) < v - 1$. Then there exists some vector $\mathbf{x}$, other than the unit vector $\mathbf{1}$ and the zero vector $\mathbf{0}$, such that $\mathbf{Ax} = \mathbf{0}$. By adding or subtracting multiples of $\mathbf{1}$, it is possible to assume that one of the elements in $\mathbf{x}$ is equal to zero while all other elements are greater than or equal to zero. Without loss of generality, let $x_1 = 0$ and $x_j \geq 0$ for $j > 1$. The first row of $\mathbf{Ax} = \mathbf{0}$ then gives

$$\sum_{j=2}^{v} a_{1j} x_j = 0 \quad \text{or} \quad \sum_{j=2}^{v} \lambda_{1j} x_j = 0$$

Now both $x_j \geq 0$ and $\lambda_{1j} \geq 0$, for all $j > 1$. If all $x_j > 0$ then $\lambda_{1j} = 0$ for all $j > 1$ which means that the first treatment is in a block of its own, hence $k = 1$. Without loss of generality, therefore, let $x_2 = x_3 = \ldots = x_{m-1} = 0$ where $m - 1 < v$, since not all x_j can be zero as $\mathbf{x}$ is not a multiple of the unit vector. Then, it follows that $\lambda_{1m} = \ldots = \lambda_{1v} = 0$. Now the lth row of $\mathbf{Ax} = \mathbf{0}$ gives, for $l = 2, 3, \ldots, m - 1$,

$$\sum_{j=m}^{v} \lambda_{lj} x_j = 0 \quad \text{which implies} \quad \lambda_{lj} = 0 \quad \text{for } j \geq m$$

This means that the 1st, 2nd,..., $(m-1)$th treatments do not occur in any block with the remaining treatments. Such a design is said to be *disconnected*. More generally, a design is disconnected if the treatments can be split into groups such that no treatment from one group occurs in any block with any treatment from a different group. A design which is not disconnected is said to be *connected*. It follows that if rank$(\mathbf{A}) < v - 1$ then the design is disconnected. If the design is connected then rank$(\mathbf{A}) \geq v - 1$ which implies that rank$(\mathbf{A}) = v - 1$. *

Although this proof refers to binary, equal block size and equal replicate designs the concept of connectedness applies equally well for non-binary designs and for designs with unequal block sizes and unequal replication. That is, *any* block design is disconnected if the blocks can be split into groups in such a way that the treatments in any one group of blocks are distinct from the treatments in the other groups. A design is connected if it is not disconnected.

Connectedness is important because in connected designs every treatment contrast is estimable from comparisons within blocks. The linear function $\mathbf{c}'\boldsymbol{\tau}$ is a *contrast* in the treatment parameters if the elements of the coefficient vector $\mathbf{c}$ sum to zero, i.e. $\mathbf{c}'\mathbf{1} = 0$. It is necessary to show, therefore, that all such contrast vectors $\mathbf{c}$ satisfy the estimability condition $\mathbf{c}' = \mathbf{c}'\boldsymbol{\Omega}\mathbf{A}$ given in Section 1.5.

A generalized inverse $\boldsymbol{\Omega}$ of the information matrix $\mathbf{A}$ has been given by the Moore–Penrose inverse $\mathbf{A}^+$ in (1.19). For a connected design, it follows immediately from (A.22) that $\mathbf{c}' = \mathbf{c}'\mathbf{A}^+\mathbf{A}$ if and only if $\mathbf{c}'\mathbf{1} = 0$. That is, in a connected design, all contrasts in the treatment parameters are estimable.

For connected designs an alternative generalized inverse, given by Shah (1959), is

$$\boldsymbol{\Omega} = (\mathbf{A} + a\mathbf{J})^{-1}, \quad a \neq 0 \tag{1.45}$$

This can be verified using (A.19) in the Appendix. With an appropriate choice of the constant a, this generalized inverse

---

* This proof is due to Dr R.G. Jones, Australian National University, Canberra.

provides a simple method of obtaining a solution to the reduced
normal equations (1.13) for many different types of designs.
For example, with $a = v^{-1}$, another generalized inverse of the
information matrix $\mathbf{A}$ of a balanced incomplete block design given
by (1.39) is $\mathbf{\Omega} = (k/\lambda v)\mathbf{I}$. More generally, John (1965) showed that
$\mathbf{\Omega} = (\mathbf{A} + a\mathbf{hh}')^{-1}$ where $a \neq 0$ and $\mathbf{h}$ is any vector such that the
columns of $\mathbf{A}$ together with $\mathbf{h}$ span a v dimensional space.

Note that if $\mathbf{\Omega}$ is obtained from (1.45) with any $a \neq 0$ or from the
Moore–Penrose generalized inverse $\mathbf{A}^+$ (the $a = 0$ case), the same
solution will be obtained to the normal equations. This is because

$$\hat{\tau} = \mathbf{\Omega q} = \sum_{i=1}^{v-1} \theta_i^{-1}\mathbf{p}_i\mathbf{p}_i'\mathbf{q} + (1/a)\mathbf{p}_v\mathbf{p}_v'\mathbf{q} = \sum_{i=1}^{v-1} \theta_i^{-1}\mathbf{p}_i\mathbf{p}_i'\mathbf{q}$$

since $\mathbf{p}_v = v^{-1/2}\mathbf{1}$ and $\mathbf{1}'\mathbf{q} = 0$. This solution is independent of
the constant a. With other generalized inverses, however, different
solutions may be obtained although estimates of treatment
contrasts will not differ since they are estimable.

Example 1.3 (continued)
From Section 1.4, the information matrix $\mathbf{A}$ and the vector of
adjusted treatment totals $\mathbf{q}$ are

$$\mathbf{A} = \frac{5}{6}\begin{pmatrix} 2 & -1 & -1 \\ -1 & 2 & -1 \\ -1 & -1 & 2 \end{pmatrix}, \quad \mathbf{q} = \frac{1}{6}\begin{pmatrix} -19 \\ 23 \\ -4 \end{pmatrix}$$

A generalized inverse of $\mathbf{A}$, using (1.45) with $a = 5/6$, is given by
$\mathbf{\Omega} = (6/15)\mathbf{I}$. Hence

$$\hat{\tau} = \mathbf{\Omega q} = \frac{1}{15}\begin{pmatrix} -19 \\ 23 \\ -4 \end{pmatrix}$$

as before. To illustrate the final point made above consider the
following generalized inverse of $\mathbf{A}$ which does not belong to the
class given by (1.45), namely

$$\mathbf{\Omega} = \frac{6}{15}\begin{pmatrix} 2 & 1 & 0 \\ 1 & 2 & 0 \\ 0 & 0 & 0 \end{pmatrix}$$

This leads to the different solution

$$\hat{\tau} = \frac{1}{15}\begin{pmatrix} -15 \\ 27 \\ 0 \end{pmatrix}$$

Note, however, that $\hat{\tau}_1 - \hat{\tau}_2 = -2.80$ and $\hat{\tau}_1 - \hat{\tau}_3 = -1.00$ for both of the above solutions. This will be the case for any contrasts in these estimates.

1.9 Orthogonality

In the linear model $y = W_1\alpha_1 + W_2\alpha_2 + \varepsilon$ the parameter vectors α_1 and α_2 are said to be *orthogonal* to each other if their least squares estimators are independent, in the sense that the estimator of α_i in this model is the same as that obtained from the model $y = W_i\alpha_i + \varepsilon$ $(i = 1, 2)$. That is, the presence or absence of one of the parameter vectors in the model does not affect the estimator of the other vector. The analysis of such orthogonal models is consequently simplified. From a consideration of the normal equations, it is clear that α_1 and α_2 will be orthogonal if and only if $W_1'W_2 = 0$. The requirement for orthogonality in block designs will now be examined.

The block and treatment parameter vectors will be orthogonal to the mean parameter μ in model (1.5) if X and Z are replaced by $X^* = X - (1/n)\mathbf{1}r'$ and $Z^* = Z - (1/n)\mathbf{1}k'$ respectively. This ensures that $X^{*'}\mathbf{1} = 0$ and $Z^{*'}\mathbf{1} = 0$. For orthogonality between blocks and treatments it is then necessary that $X^{*'}Z^* = 0$. Simplifying this expression, using (1.6)–(1.8), the condition for orthogonality in a block design is seen to reduce to

$$N = (1/n)rk' \tag{1.46}$$

With incidence matrix N given by (1.46), the information matrix A and vector of adjusted treatment totals q, given in (1.14) and (1.15) respectively, become

$$A = r^\delta - (1/n)rr' \tag{1.47}$$

$$q = T - \bar{y}r \tag{1.48}$$

It is easily verified, by showing that $A\Omega A = A$, that a generalized inverse of A is given by

$$\Omega = r^{-\delta} \tag{1.49}$$

Hence

$$\hat{\tau} = \Omega q = r^\delta T - \bar{y}\mathbf{1} \tag{1.50}$$

Contrasts in the treatment parameters are estimable with estimates given by contrasts in the unadjusted treatment means. That is, $c'\tau$ is estimable if and only if $c'\mathbf{1} = 0$, with estimator $c'(r^{-\delta}T)$.

The adjusted sum of squares due to treatment effects is

$$S(\tau/\mu, \beta) = \hat{\tau}'\mathbf{q} = \mathbf{T}'\mathbf{r}^{-\delta}\mathbf{T} - G^2/n$$

which is the same as the unadjusted treatment sum of squares given in (1.35). Thus, for orthogonal block designs,

$$S(\tau/\mu, \beta) = S(\tau/\mu) \qquad (1.51)$$

In a similar way it can be shown that $S(\beta/\mu, \tau) = S(\beta/\mu)$.

Orthogonality between treatments and blocks has resulted in a considerable simplification in the analysis. Estimates of treatment parameters are based on treatment means, no adjustment for block totals is necessary. It is also unnecessary to eliminate block effects in calculating the treatment sum of squares. As a consequence, all the information on treatment effects is contained in comparisons within blocks. Similarly, block parameters can be estimated and tested without the need to eliminate treatment effects.

The construction of orthogonal designs follows immediately from (1.46) since knowledge of the incidence matrix $\mathbf{N}$ gives the design directly; remember that the (ij)th element of $\mathbf{N}$ gives the number of times the ith treatment occurs in the jth block.

Example 1.5
An orthogonal design for $v = 2$ treatments with $\mathbf{r} = (3, 3)'$, $\mathbf{k} = (2, 4)'$ and $n = \mathbf{r}'\mathbf{1} = \mathbf{k}'\mathbf{1} = 6$ has, from (1.46),

$$\mathbf{N} = \begin{pmatrix} 1 & 2 \\ 1 & 2 \end{pmatrix}$$

Thus, both treatments 0 and 1 occur once in the first block and twice in the second block. The two blocks of this design are, therefore, (0 1) and (0 0 1 1).

Of particular importance are orthogonal designs for equal block sizes and equal treatment replication. It follows from (1.46) that $\mathbf{N}$ must a multiple of $\mathbf{J}$, i.e. each treatment must occur the same number of times in each block. The complete block design is, therefore, an orthogonal design, and consequently has a simple analysis, as was shown in Section 1.1.

CHAPTER 2

Efficiency factors

2.1 Introduction

In an orthogonal design all the information on the treatments is obtained from the intra-block analysis. That is, contrasts amongst the treatment parameters are estimated entirely from comparisons made within blocks. In a non-orthogonal design some of the information is obtained from these within-block comparisons, but the remaining information has to be recovered, using an inter-block analysis, from comparison between blocks. Different designs of the same size can be assessed, therefore, on the basis of the amount of information available from both within and between blocks. For this purpose *efficiency factors*, obtained from the intra-block analysis, will be defined and used to provide criteria for comparing different non-orthogonal block designs.

Attention will be mainly restricted to designs in which the v treatments are set out in b blocks each of size k, such that each treatment is replicated r times. The efficiency factors of a block design are then obtained from a comparison of the variances of estimated treatment contrasts in the design with those in an orthogonal design using the same number of experimental units, under the assumption that the error variance σ^2 is the same for both designs. This is equivalent to making comparisons against the design that would have been obtained if no blocking had been used. The case of unequal replication will be discussed in Section 2.6.

The concept of an efficiency factor should not be confused with the measure of the gain (or loss) resulting from the use of a block design or of a design of a given size, the so-called *efficiency* of a design. The purpose of blocking is to reduce the error variance σ^2 by exercising greater local control on the experiment; the gain in efficiency resulting from blocking will often be substantial. There will also frequently be a gain in efficiency resulting from using designs with small rather than large block sizes. However, designs of the same size will often differ in the precision with which treatment effects are estimated; it is the purpose of efficiency factors to provide convenient measures of these differences.

2.2 Pairwise efficiency factors

In a connected design all pairwise comparisons $\tau_i - \tau_j$ $(i \neq j)$ are estimable with estimators $\hat{\tau}_i - \hat{\tau}_j$, where $\hat{\tau}_i$ is the least squares estimator of τ_i obtained from (1.17). Further, from (1.26)

$$\mathrm{var}(\hat{\tau}_i - \hat{\tau}_j) = (\omega_{ii} + \omega_{jj} - 2\omega_{ij})\sigma^2 = \nu_{ij}\sigma^2 \quad (i \neq j),$$

where ω_{ij} is the (ij)th element of a generalized inverse $\mathbf{\Omega}$ of the information matrix $\mathbf{A}$ given in (1.14). For a complete block design this variance is, from (1.49), $2\sigma^2/r$ for all $i \neq j$. The efficiency factor e_{ij} for the pairwise comparison $\tau_i - \tau_j$ is then defined to be

$$e_{ij} = \frac{2\sigma^2/r}{\nu_{ij}\sigma^2} = \frac{2}{r\nu_{ij}} \tag{2.1}$$

The average variance of all estimated pairwise comparisons is

$$\bar{\nu} = \frac{2\sigma^2}{v(v-1)} \sum_{i>j}\sum \nu_{ij}$$

so that an overall average efficiency factor can be defined by

$$E = \frac{2\sigma^2/r}{\bar{\nu}} = \frac{v(v-1)}{r \sum_{i>j}\sum \nu_{ij}} \tag{2.2}$$

Note that E is also the harmonic mean of the pairwise efficiency factors e_{ij} given in (2.1).

In an experiment it is often of interest to look for differences between pairs of treatments. Hence the pairwise efficiency factors e_{ij} are useful measures against which to judge the suitability of different designs. The distribution of the e_{ij} values, or their range or minimum (or maximum) value, give important information relating to the usefulness of any design. If all differences are of equal importance then the average efficiency factor E provides a convenient summary of the e_{ij}.

2.3 Canonical efficiency factors

The information matrix $\mathbf{A}$ has a complete set of orthogonal eigenvectors since it is a symmetric matrix. If there are multiplicities among the eigenvalues then the eigenvectors will not be uniquely determined, although a set of linearly independent eigenvectors can be found which will span the row or column space of $\mathbf{A}$. Further, the eigenvectors of $\mathbf{A}$ can be normalized so that the sum of squares of the elements of any eigenvector is unity.

Let $\theta_1, \theta_2, \ldots, \theta_v$ be the eigenvalues of $\mathbf{A}$, not necessarily all distinct, with corresponding eigenvectors $\mathbf{p}_1, \mathbf{p}_2, \ldots, \mathbf{p}_v$ satisfying

$$\mathbf{p}_i'\mathbf{p}_j = \begin{cases} 1, & i = j \\ 0, & i \neq j \end{cases} \tag{2.3}$$

Then, as previously given in (1.18), $\mathbf{A}$ can be expressed in canonical form as

$$\mathbf{A} = \sum_{i=1}^{v} \theta_i \mathbf{p}_i \mathbf{p}_i' \tag{2.4}$$

Since $\mathbf{A}$ is a singular matrix at least one eigenvalue, θ_v say, will be zero with corresponding eigenvector $\mathbf{p}_v = (v^{-1/2})\mathbf{1}$. For a connected design all other $v - 1$ eigenvalues are non-zero, so that a generalized inverse of $\mathbf{A}$ is then given by

$$\mathbf{A}^{+} = \sum_{i=1}^{v-1} \theta_i^{-1} \mathbf{p}_i \mathbf{p}_i' \tag{2.5}$$

Attention will again be restricted to connected designs; for disconnected designs the main interest would in any case centre on identifying the non-estimable treatment contrasts rather than on any notion of efficiency factors.

Since $\mathbf{p}_i'\mathbf{p}_v = 0$ $(i \neq v)$, it follows that $\mathbf{p}_i'\tau$ represents an estimable contrast in the treatment parameters. Its estimator is

$$\mathbf{p}_i'\hat{\tau} = \theta_i^{-1} \mathbf{p}_i'\mathbf{q}$$

where $\mathbf{q}$ is the vector of adjusted treatment totals given in (1.15). Further, using (1.26),

$$\mathrm{var}(\mathbf{p}_i'\hat{\tau}) = \mathbf{p}_i'\mathbf{A}^{+}\mathbf{p}_i\sigma^2 = \sigma^2/\theta_i$$

For a complete block design this variance is σ^2/r, using (1.49). Hence, the efficiency factor e_i of the contrast $\mathbf{p}_i'\tau$ is

$$e_i = \theta_i/r \quad (i = 1, 2, \ldots, v - 1) \tag{2.6}$$

Note that the e_i in (2.6) are the non-zero eigenvalues of the matrix

$$\mathbf{A}^{*} = (1/r)\mathbf{A} = \mathbf{I} - (1/rk)\mathbf{N}\mathbf{N}' \tag{2.7}$$

The contrasts $\mathbf{p}_i'\tau$ have been called the *basic contrasts* of a design by Pearce, Calinski and Marshall (1974). The non-zero eigenvalues of the matrix $\mathbf{A}^*$ in (2.7) are the efficiency factors of the basic contrasts and have been called the *canonical efficiency factors* by James and Wilkinson (1971).

The adjusted treatment sum of squares (1.31) can now be written as

$$\hat{\tau}'\mathbf{q} = \mathbf{q}'\mathbf{A}^+\mathbf{q} = (1/r) \sum_{i=1}^{v-1} e_i^{-1}(\mathbf{p}_i'\mathbf{q})^2 \qquad (2.8)$$

This means that the basic contrasts provide an orthogonal decomposition of the $v - 1$ degrees of freedom for treatments into single degree of freedom components. If there are multiplicities amongst the canonical efficiency factors then this decomposition will not be unique. In particular, if all canonical efficiency factors are equal, as in a complete block design and a balanced incomplete block design, then any set of $v - 1$ linearly independent orthogonal contrasts will provide a decomposition of the treatment sum of squares.

Two important properties of canonical efficiency factors will now be considered. The first is that their harmonic mean is equal to the average efficiency factor E based on pairwise treatment differences given in (2.2). Let $\mathbf{d}_{jk}$ be the vector with jth element equal to 1, the kth element equal to -1 and the remaining $v - 2$ elements equal to zero. Then

$$\text{var}(\hat{\tau}_j - \hat{\tau}_k) = \mathbf{d}_{jk}'\mathbf{A}^+\mathbf{d}_{jk}\sigma^2 = (\sigma^2/r) \sum_{i=1}^{v-1} e_i^{-1}(\mathbf{d}_{jk}'\mathbf{p}_i)^2$$

The average variance over all pairwise differences is, with $w = v(v - 1)/2$,

$$
\begin{aligned}
\bar{\nu} &= (1/w) \sum_{j<k} \sum \text{var}(\hat{\tau}_j - \hat{\tau}_k) \\
&= (\sigma^2/rw) \sum_i e_i^{-1} \Big\{ \sum_{j<k} \sum (\mathbf{d}_{jk}'\mathbf{p}_i)^2 \Big\} \\
&= (\sigma^2/rw) \sum_i e_i^{-1} \{ v\mathbf{p}_i'\mathbf{p}_i - (\mathbf{p}_i'\mathbf{1})^2 \} \\
&= (2\sigma^2/r) \Big\{ \sum_{i=1}^{v-1} e_i^{-1}/(v - 1) \Big\}
\end{aligned}
$$

The average efficiency factor E is then given by

$$E = (v-1)/\sum_{i=1}^{v-1} e_i^{-1} \qquad (2.9)$$

namely the harmonic mean of the canonical efficiency factors.

The second property is that no treatment contrast can have an efficiency factor smaller (or larger) than the smallest (or largest) canonical efficiency factor. Any contrast vector $\mathbf{d}$ can be written as $\mathbf{d} = \sum a_i \mathbf{p}_i$ for some constants $a_1, a_2, \ldots, a_{v-1}$. Then

$$\text{var}(\mathbf{d}'\hat{\boldsymbol{\tau}}) = (\sigma^2/r)\sum(a_i^2/e_i)$$

Since for a complete block design $e_i = 1$ $(i = 1, 2, \ldots, v-1)$, the efficiency factor of the contrast $\mathbf{d}'\boldsymbol{\tau}$ is

$$E(\mathbf{d}'\boldsymbol{\tau}) = \frac{\sum a_i^2}{\sum(a_i^2/e_i)}$$

Let e_{min} and e_{max} be the smallest and largest canonical efficiency factors respectively. Then, as

$$e_{max}^{-1}\sum a_i^2 \le \sum(a_i^2/e_i) \le e_{min}^{-1}\sum a_i^2$$

it follows that

$$e_{min} \le E(\mathbf{d}'\boldsymbol{\tau}) \le e_{max} \qquad (2.10)$$

Canonical efficiency factors provide a useful summary of the properties of a design. They are particularly important if the basic contrasts correspond to the contrasts of interest in the experiment, as is frequently the case in the multiple replicate factorial designs discussed in Chapter 9. The use of these factors to define optimality criteria is discussed in the next section.

2.4 Optimality criteria

Canonical efficiency factors provide a measure of the amount of information available on basic treatment contrasts from within-block comparisons. Since comparisons made within blocks are usually more accurately made than those between blocks, the aim in selecting a design, assuming all treatments to be of equal interest, should be to choose one which has these efficiency factors as large as possible. Various optimality criteria have been proposed.

If the harmonic mean of the canonical efficiency factors, or the average efficiency factor of pairwise treatment factors, of a design is at least as large as that of any other design then the design is said to be *A-optimal*. In view of (2.10), a criterion based on the smallest canonical efficiency factor is of interest since it indicates the worst that a design can do in estimating treatment contrasts. A design whose smallest canonical efficiency factor is at least as large as that of any other design is said to be *E-optimal*. A further criterion uses the geometric mean (or product) of the canonical efficiency factors. A design whose geometric mean efficiency factor is at least as large as that of any other design is said to be *D-optimal*. D-optimality is a criterion which has proved useful in the context of regression analysis, where it has an interpretation in terms of its equivalence with minimizing the maximum variance of predicted responses; such an interpretation is less meaningful for block designs.

A design which is optimal under any one of the above criteria is not necessarily optimal under the others; although John and Williams (1982) have conjectured that if a block design is A-optimal then it is also D-optimal. However, evidence gained from studies of different types of designs suggests that a design which is optimal or performs well on one criterion tends to perform well on the other criteria.

Another useful criterion is that of *(M,S)-optimality*, introduced by Shah (1960) and Eccleston and Hedayat (1974). It is a two-stage criterion. Firstly, a class of designs which maximizes the mean of the efficiency factors is formed (M-optimality). Then, within this class, the design (or designs) which minimizes the spread or variance of the efficiency factors is identified (S-optimality). That is, $\sum e_i$ is first maximized and then $\sum e_i^2$ minimized. This criterion, therefore, uses the intuitively appealing requirement that all the efficiency factors should be as equal as possible.

If the canonical efficiency factors of a design are all equal then the design is clearly optimal under all four optimality criteria given above. Such a design is said to be *efficiency-balanced*; a complete block design is an example of such a design. From (2.4) it follows that for such designs

$$\mathbf{A} = rE \sum_{i=1}^{v-1} \mathbf{p}_i \mathbf{p}_i'$$

Now $\sum_{i=1}^{v} \mathbf{p}_i \mathbf{p}_i' = \mathbf{I}$, as it is the sum of v mutually orthogonal idempotent matrices, so that

$$\mathbf{A} = rE(\mathbf{I} - \mathbf{K}) \tag{2.11}$$

since $\mathbf{p}_v = (v^{-1/2})\mathbf{1}$. This matrix has only one distinct non-zero eigenvalue so that a design is efficiency-balanced if and only if the information matrix is of the form given by (2.11). It follows from (1.39) that a balanced incomplete block design is also efficiency-balanced, with

$$E = \lambda v/rk \qquad (2.12)$$

If $\mathbf{A}$ is given by (2.11) then from (1.19),

$$\mathbf{A}^+ = (1/rE)(\mathbf{I} - \mathbf{K})$$

and so the design will also be *variance-balanced* in the sense that the variances of all estimated pairwise treatment differences will be the same, and equal to

$$\mathrm{var}(\hat{\tau}_i - \hat{\tau}_j) = \frac{2\sigma^2}{rE} \quad \text{for all } i \neq j \qquad (2.13)$$

An equally replicated variance-balanced design is also efficiency-balanced. Williams (1975b) shows that this is not the case for unequally replicated designs with $v > 2$. This point is discussed further in Section 2.6.

A number of computer algorithms are available that construct optimal or near optimal block and row–column designs. They all use some form of interchange procedure whereby pairs of treatments are interchanged in the design if an improvement results in the chosen optimality criterion. Interchanges continue until no further improvements can be made. Some arbitrary starting design is chosen to begin the search. A drawback with such a procedure is that it can become trapped at a local optimum. Two approaches have been used to overcome this difficulty. The first is to make repeated runs of the algorithm using different starting designs, and then choosing the best design over all runs; this method has been used by Nguyen and Williams (1993a). The second uses simulated annealing, whereby randomly chosen interchanges are not only accepted if they result in an improvement in the chosen optimality criterion, but are accepted with some low probability even if no improvement results; algorithms that are based on simulated annealing have been described by John and Whitaker (1993) and Venables and Eccleston (1993). All of these algorithms produce efficient designs and can handle large numbers of treatments. Interchange algorithms are discussed in detail in Section 6.4.

The optimality criteria given above are appropriate when all treatments are of equal interest. However, experiments are often carried out in which some treatment contrasts are of more importance than others. Different criteria will, therefore, be needed for choosing designs appropriate for such experiments. Pearce (1974) proposed minimizing the weighted mean of the efficiency factors of interest. Freeman (1976a) suggested minimizing the weighted mean given by $\sum w_i \, \mathrm{var}(\mathbf{c}_i' \hat{\tau})$ where $\mathbf{c}_i' \tau$ represents a contrast of interest and w_i is the weight to be attached to this contrast $(i = 1, 2, \ldots, t$, say$)$. If $\mathbf{C}$ is the $v \times t$ matrix whose ith column is $\mathbf{c}_i$ and $\mathbf{w}$ is the weighting vector with ith element w_i, then the criterion is one of minimizing

$$\mathrm{trace}(\mathbf{w}^\delta \mathbf{C}' \mathbf{A}^+ \mathbf{C}) = \mathrm{trace}(\mathbf{A}^+ \mathbf{C} \mathbf{w}^\delta \mathbf{C}') \qquad (2.14)$$

The difficulty, of course, is choosing appropriate weights as the choice in practice is most likely to be a highly subjective one. Jones (1976) has given details of a computer algorithm to derive optimal block designs using the criterion given in (2.14).

2.5 Simple counting rules for producing near-optimal designs

The search for optimal designs in a large family of designs of a given size invariably involves a mixture of theory and computing. The use of theoretical results can often limit the number of possible designs that have to be considered. Even so there will often be large numbers of designs for which the canonical efficiency factors have to be computed. Eigenvalue computations are relatively time consuming and, consequently, attempts have been made to find criteria based on quick and simple counting rules. The idea is to use such rules to identify a small group of designs which are near optimal and then, if necessary, to choose the best designs from this group using the canonical efficiency factors and appropriate optimality criteria. That is, the counting rules are aimed at an initial sifting out of the less efficient designs.

One such rule is, in fact, given by the (M,S)-optimality criterion given in the previous section. Consider first the class of binary designs with $k \le v$, defined in Section 1.8. Using (2.7),

$$\sum e_i = \mathrm{trace}(\mathbf{A}^*) = v(k-1)/k$$

and

$$\sum e_i^2 = \mathrm{trace}(\mathbf{A}^{*2}) = v(k-1)^2/k^2 + \sum_{i \ne j} \sum \lambda_{ij}^2/(rk)^2$$

where λ_{ij} gives the number of blocks containing both the ith and jth treatments. Hence, for designs of a given size, $\sum e_i$ is a constant and $\sum e_i^2$ is minimized by minimizing the sum of squares, $\sum \sum \lambda_{ij}^2$, of the off-diagonal elements of the concurrence matrix $\mathbf{NN'}$. An (M,S)-optimal binary design is, therefore, obtained by ensuring that the treatment concurrences are as equal as possible. The rule is simple to apply and does not involve the calculation of eigenvalues. However, it still may lead to a large class of designs. Further searching can be restricted to this class. For equal replicate designs, it has been conjectured that the A-, D- and E-optimal designs are to be found in the class of (M,S)-optimal designs; see John and Williams (1982).

When $k > v$ the designs are necessarily non-binary. An (M,S)-optimal non-binary design belongs to the class of designs obtained by appending a binary design to sets of complete block designs. That is, the required design will have incidence matrix $\mathbf{N} = p\mathbf{J} + \mathbf{N_0}$ where p is some positive integer and $\mathbf{N_0}$ is the incidence matrix of the binary design. Further, if the binary design is (M,S)-optimal then so is the corresponding non-binary design.

Although minimizing $\sum \sum \lambda_{ij}^2$, or equivalently trace$\{(\mathbf{NN'})^2\}$, is a quick and simple method of finding (M,S)-optimal designs it will in general still produce a large class of designs. Further, not all these designs will be good on other optimality criteria. It becomes necessary, therefore, to apply other rules to pick out the best designs from this class. Lamacraft and Hall (1982) in constructing tables of cyclic designs used, at a second stage, the criterion of minimizing trace$\{(\mathbf{NN'})^3\}$. A rationale behind this choice, for connected designs, is as follows. Writing $\mathbf{A}^*$, given in (2.7), as

$$\mathbf{A}^* = (\mathbf{I} - \mathbf{K}) - \mathbf{W}$$

where $\mathbf{W} = (1/rk)\mathbf{NN'} - \mathbf{K}$, it follows that $\mathbf{A}^{*+}$ can be expanded as

$$\mathbf{A}^{*+} = (\mathbf{I} - \mathbf{K}) + \sum_{j=1}^{\infty} \mathbf{W}^j$$

Then, since the canonical efficiency factors are the non-zero eigenvalues of $\mathbf{A}^*$,

$$\sum_{i=1}^{v-1} e_i^{-1} = (v - 1) + \sum_{j=1}^{\infty} \text{trace}(\mathbf{W}^j) \qquad (2.15)$$

But since $\mathbf{NN'K} = rk\mathbf{K}$, it follows that the traces of

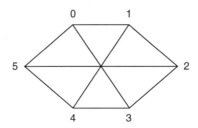

Figure 2.1. *The treatment concurrence graph for Example 2.1.*

successive powers of $\mathbf{NN'}$ provide simple criteria to use in the search for A-optimal designs. The smaller powers make the greatest contribution to the sum in (2.15), so that it becomes intuitively reasonable to minimize sequentially $\text{trace}\{(\mathbf{NN'})^2\}$, $\text{trace}\{(\mathbf{NN'})^3\}$ and so on in the search for efficient binary designs.

For non-binary designs it will also be necessary to first minimize $\text{trace}(\mathbf{NN'})$, since this value will no longer be constant. An alternative to this sequential approach is to minimize

$$f_m = \sum_{j=1}^{m} (rk)^{-j} \text{trace}\{(\mathbf{NN'})^j\} \qquad (2.16)$$

for some small value of m.

Paterson (1983) proposed a similar sequential procedure, based on the number of circuits of different length in the treatment concurrence graph of a design. This graph has been defined in Section 1.8.

Example 2.1
Consider the following design for six treatments labelled $0, 1, \ldots 5$, in nine blocks of two, where blocks are set out in columns:

$$\begin{array}{ccccccccc}
0 & 1 & 2 & 3 & 4 & 5 & 0 & 1 & 2 \\
1 & 2 & 3 & 4 & 5 & 0 & 3 & 4 & 5
\end{array}$$

Its treatment concurrence graph is shown in Fig. 2.1. Treatment 0, for instance, occurs in the same block as treatments 1, 3 and 5 but not with 2 and 4. Hence lines connect point 0 with points 1, 3 and 5 but not with points 2 and 4.

A *path* joining points i and j is a route from i to j traversing lines in the graph, while a *circuit* is a path joining a point to itself.

For example, in the graph in Fig. 2.1, 0–1–2 is a path joining points 0 and 2, and 0–1–4–5–0 is a circuit at point 0.

Each path in the graph contributes to the estimate of a pairwise treatment difference in the design. To illustrate this suppose that the observations in the first two blocks in the design of Example 2.1 are y_1, y_2, y_3 and y_4 respectively. Then, assuming the intra-block model (1.3), $y_1 - y_2$ provides an estimate of $\tau_0 - \tau_1$ with variance $2\sigma^2$, and $(y_1 - y_2) - (y_3 - y_4)$ an estimate of $\tau_0 - \tau_2$ with variance $4\sigma^2$. The first estimate corresponds to the path 0–1 of length 1, while the second corresponds to the path 0–1–2 of length 2. In general a path of length h will contribute an estimate of a treatment difference with variance $2h\sigma^2$. Paths of differing lengths give estimates which can be combined to give overall estimates of each pairwise treatment difference.

Circuits, on the other hand, contribute nothing useful to the estimation of such treatment differences, since a comparison of a treatment with itself is of no interest. Consequently, good designs might be expected to have few circuits of short lengths. If c_h is the number of circuits of length h in the treatment concurrence graph, then Paterson (1983) argues that a reasonable criterion to use in the search for good designs would be to sequentially minimize c_2, c_3, c_4 and so on.

The fact that $c_h = \text{trace}(\mathbf{B}^h)$, where $\mathbf{B}$ is the adjacency matrix of the treatment concurrence graph given by $\mathbf{B} = \mathbf{NN}' - r\mathbf{I}$, connects this approach with that based on (2.15).

2.6 Efficiency factors for unequally replicated designs

For designs with unequal treatment replication two measures of efficiency factors can be defined. Which should be used to assess different designs will depend on the constraints imposed by experimental requirements.

One measure will be appropriate when the replication vector is regarded as fixed. That is, it is a condition of the experiment that the ith treatment is replicated r_i times in the design, where r_i is fixed $(i = 1, 2, \ldots, v)$. Efficiency factors are then obtained by comparing variances of estimated treatment contrasts with those obtained from an orthogonal design, i.e. with

$$\text{var}(\mathbf{c}'\hat{\tau}) = \mathbf{c}'\mathbf{r}^{-\delta}\mathbf{c}\sigma^2 \qquad (2.17)$$

This provides a measure of the effectiveness of blocking since (2.17) gives the variances that would have been obtained (assuming the same σ^2) if blocks had been ignored in the design.

The second measure will be more suitable when the total number of experimental units is fixed, i.e. when $n = \mathbf{r}'\mathbf{1}$ is fixed but not the individual elements of the vector $\mathbf{r}$. A basis for the comparison of designs will then be the average replication $\bar{r} = n/v$. Hence, for connected designs, if $\theta_1, \theta_2, \ldots, \theta_{v-1}$ are the non-zero eigenvalues of the information matrix $\mathbf{A}$ given in (1.14) then, following (2.6), corresponding efficiency factors are given by $\theta_j/\bar{r}$ $(j = 1, 2, \ldots, v - 1)$. Criteria based on these efficiency factors can then be used to assess different designs. In general, when treatments are of equal interest, optimal designs will have the property that the treatments are equally replicated; Jones and Eccleston (1980), however, give three examples with block size 2 where the A-optimal designs have unequal rather than equal replication. Of course, an equally replicated design will not be available if $\bar{r}$ is not an integer.

Since this second measure is relatively straightforward, the remainder of this section will be concerned with the case where $\mathbf{r}$ is regarded as fixed. Following Ceranka and Mejza (1979), the concepts of a basic contrast and a canonical efficiency factor can be generalized. Let e_j be an eigenvalue corresponding to an eigenvector $\mathbf{s}_j$ of the matrix $\mathbf{A}$ given in (1.14) with respect to $\mathbf{r}^\delta$, i.e. $\mathbf{As}_j = e_j\mathbf{r}^\delta\mathbf{s}_j$ $(j = 1, 2, \ldots, v)$. The eigenvectors $\mathbf{s}_j$ can be chosen to satisfy

$$\mathbf{s}_i\mathbf{r}^\delta\mathbf{s}_j = \left\{ \begin{array}{ll} 1, & i = j \\ 0, & i \neq j \end{array} \right. \tag{2.18}$$

Again, since $\mathbf{A}$ is a singular matrix, at least one eigenvalue, e_v say, is zero with corresponding eigenvector $\mathbf{s}_v = (n^{-1/2})\mathbf{1}$. For a connected design all other $v - 1$ eigenvalues are non-zero. Now if $\mathbf{p}_j = \mathbf{r}^\delta\mathbf{s}_j$ $(j = 1, 2, \ldots, v - 1)$ then $\mathbf{p}'_j\tau$ is a contrast in the treatment parameters. Further, using (1.26) and the above results,

$$\begin{array}{rcl} \text{var}(\mathbf{p}'_j\hat{\tau}) & = & \mathbf{p}'_j\mathbf{\Omega}\mathbf{p}_j\sigma^2 \\ & = & \mathbf{s}'_j\mathbf{r}^\delta\mathbf{\Omega}\mathbf{r}^\delta\mathbf{s}_j\sigma^2 \\ & = & e_j^{-2}\mathbf{s}'_j\mathbf{A}\mathbf{\Omega}\mathbf{A}\mathbf{s}_j\sigma^2 \\ & = & e_j^{-2}\mathbf{s}'_j\mathbf{A}\mathbf{s}_j\sigma^2 \\ & = & e_j^{-1}\sigma^2 \end{array}$$

For a complete block design, with $\mathbf{\Omega} = \mathbf{r}^{-\delta}$,

$$\text{var}(\mathbf{p}'_j\hat{\tau}) = \mathbf{s}'_j\mathbf{r}^\delta\mathbf{r}^{-\delta}\mathbf{r}^\delta\mathbf{s}_j = \mathbf{s}'_j\mathbf{r}^\delta\mathbf{s}_j\sigma^2 = \sigma^2$$

Hence, the efficiency factor of the contrast $\mathbf{p}'_j\tau$ is e_j. These contrasts are the basic contrasts of the design and the e_j are the canonical efficiency factors. Extending (2.7) to the unequal replicate case, the canonical efficiency factors are more conveniently obtained as the eigenvalues of the symmetric matrix

$$\mathbf{A}^* = \mathbf{r}^{-\delta/2} \mathbf{A} \mathbf{r}^{-\delta/2} = \mathbf{I} - \mathbf{r}^{-\delta/2} \mathbf{N} \mathbf{k}^{-\delta} \mathbf{N}' \mathbf{r}^{-\delta/2} \qquad (2.19)$$

where $\mathbf{r}^{-\delta/2}$ is the inverse of $\mathbf{r}^{\delta/2}$, which is the diagonal matrix with ith diagonal element equal to $r_i^{1/2}$.

With unequal replicate designs, therefore, optimality criteria based on the eigenvalues of the information matrix $\mathbf{A}$ or the matrix $\mathbf{A}^*$ in (2.19) can be used depending on whether the purpose is to compare designs where only the total size of the experiment is fixed or where the replication vector is fixed.

2.7 Duality

The dual design of a block design with incidence matrix $\mathbf{N}$ is the block design with incidence matrix $\mathbf{N}'$. Thus the dual is obtained by interchanging the treatment and block symbols in the original design.

Example 2.2
Consider the following binary design for $v = 6, r = 2, k = 4$ and $b = 3$, where again blocks are written in columns:

$$
\begin{array}{ccc}
0 & 0 & 1 \\
1 & 2 & 2 \\
3 & 3 & 4 \\
4 & 5 & 5
\end{array}
$$

The incidence matrix $\mathbf{N}$, with a row for each treatment and a column for each block, is

$$
\mathbf{N} =
\begin{pmatrix}
1 & 1 & 0 \\
1 & 0 & 1 \\
0 & 1 & 1 \\
1 & 1 & 0 \\
1 & 0 & 1 \\
0 & 1 & 1
\end{pmatrix}
$$

Consequently, the incidence matrix of the dual design is

$$
\mathbf{N}' =
\begin{pmatrix}
1 & 1 & 0 & 1 & 1 & 0 \\
1 & 0 & 1 & 1 & 0 & 1 \\
0 & 1 & 1 & 0 & 1 & 1
\end{pmatrix}
$$

Hence, the dual design is

$$
\begin{array}{cccccc}
0 & 0 & 1 & 0 & 0 & 1 \\
1 & 2 & 2 & 1 & 2 & 2
\end{array}
$$

and has parameters $v_d = b = 3$, $r_d = k = 4$, $k_d = r = 2$ and $b_d = v = 6$.

Two useful relationships between a design and its dual can now be established. One concerns the canonical efficiency factors and their harmonic mean, and the other the generalized inverse of the information matrix. It will be assumed, for the first relationship, that the original design and, hence, its dual design are connected.

The average efficiency factor E_d, say, of the dual design is the harmonic mean of the non-zero eigenvalues of the $b \times b$ matrix

$$
\mathbf{A}_d^* = \mathbf{I} - \mathbf{k}^{-\delta/2} \mathbf{N}' \mathbf{r}^{-\delta} \mathbf{N} \mathbf{k}^{-\delta/2} \tag{2.20}
$$

This follows from the form of $\mathbf{A}^*$ given in (2.19). Note that

$$
\mathbf{A}^* = \mathbf{I} - \mathbf{B}\mathbf{B}', \qquad \mathbf{A}_d^* = \mathbf{I} - \mathbf{B}'\mathbf{B}
$$

where $\mathbf{B} = \mathbf{r}^{-\delta/2} \mathbf{N} \mathbf{k}^{-\delta/2}$.

Now consider the case where $v > b$. $\mathbf{A}_d^*$ then has a single zero eigenvalue and $(b-1)$ non-zero eigenvalues $e_1, e_2, \ldots, e_{b-1}$. Thus

$$
E_d = \frac{b-1}{\sum_{i=1}^{b-1} e_i^{-1}} \tag{2.21}
$$

As the matrices $\mathbf{B}\mathbf{B}'$ and $\mathbf{B}'\mathbf{B}$ have the same non-zero eigenvalues, the eigenvalues of the matrix $\mathbf{A}^*$ of (2.19) consist of the b eigenvalues of $\mathbf{A}_d^*$ and $(v-b)$ eigenvalues equal to 1. Hence, the average efficiency factor of the original design is

$$
E = \frac{v-1}{(v-b) + \sum_{i=1}^{b-1} e_i^{-1}}
$$

Using (2.21) gives

$$
E = \frac{v-1}{(v-b) + (b-1)E_d^{-1}} \tag{2.22}
$$

If $v \le b$, then $\mathbf{A}_d^*$ has a single zero eigenvalue, $(v-1)$ non-zero eigenvalues $e_1, e_2, \ldots, e_{v-1}$ and $(b-v)$ eigenvalues equal to 1. Hence

$$
E_d = \frac{b-1}{(b-v) + \sum_{i=1}^{v-1} e_i^{-1}} \tag{2.23}
$$

Since the non-zero eigenvalues of $\mathbf{A}^*$ are now $e_1, e_2, \ldots, e_{v-1}$, it follows from (2.23) that E is again given by (2.22). This establishes a relationship between the average efficiency factors of a design and its dual. Note that E increases with E_d.

These results are useful in a number of ways. If v is very much greater than b then the canonical efficiency factors and the average efficiency factor are most easily obtained from the dual design. If it is known that a design has a high average efficiency factor then, from (2.22), so will its dual. For example, the dual design with three treatments in Example 2.2 is a balanced incomplete block design with $\lambda = 2$ and, from (2.12), $E_d = 3/4$. It follows that the design with six treatments in blocks of four is A-optimal with $E = 15/17$, using (2.22). This design is also D- and E-optimal.

The second relationship between a design and its dual concerns the generalized inverse of the information matrix. For the dual design the information matrix becomes

$$\mathbf{A}_d = \mathbf{k}^\delta - \mathbf{N}'\mathbf{r}^{-\delta}\mathbf{N} \qquad (2.24)$$

If $\mathbf{\Omega}_d$ is a generalized inverse of this matrix then it can be established that

$$\mathbf{\Omega} = \mathbf{r}^{-\delta}(\mathbf{r}^\delta + \mathbf{N}\mathbf{\Omega}_d\mathbf{N}')\,\mathbf{r}^{-\delta} \qquad (2.25)$$

is a generalized inverse of the information matrix $\mathbf{A}$ by showing that $\mathbf{A}\mathbf{\Omega}\mathbf{A} = \mathbf{A}$.

If $\mathbf{\Omega}_d$ is easier to obtain than $\mathbf{\Omega}$, which would often be the case, for instance, if v is very much greater than b, then $\mathbf{\Omega}$ can be conveniently obtained by substituting $\mathbf{\Omega}_d$ into (2.25).

2.8 Upper bounds for the average efficiency factor

In the search for A-optimal designs it is useful to have available an upper bound for the average efficiency factor E as a measure against which to assess the scope for possible improvement. For instance, a computer search could stop when a design is found which is sufficiently close to the bound. A number of upper bounds have been proposed and, for a given set of parameters, the best bound will be provided by the minimum value of those available.

Binary designs, having $v \geq k$, will be considered first. Since the canonical efficiency factors are all positive, the harmonic mean cannot exceed the arithmetic mean $\bar{e}$ of these factors. Hence, the simplest upper bound for the average efficiency factor E is $\bar{e} = \text{trace}(\mathbf{A}^*)/(v - 1)$, where $\mathbf{A}^*$ is given by (2.7). For binary designs, $\text{trace}(\mathbf{A}^*) = v(k-1)/k$ so that the bound is

$$\bar{e} = \frac{v(k-1)}{k(v-1)} \tag{2.26}$$

However, $E = \bar{e}$ only when the canonical efficiency factors of the design are all equal, i.e. when the design is efficiency-balanced. Improved bounds can be obtained for those cases where efficiency-balanced designs cannot be constructed. Jarrett (1977) gives the bound

$$U_1 = \bar{e} - \frac{(v-2)S^2}{\bar{e} + (v-3)S} \tag{2.27}$$

where

$$(v-1)(v-2)S^2 = S_2$$

and where S_2 is the corrected second moment of the $v-1$ canonical efficiency factors given by

$$S_2 = \sum (e_i - \bar{e})^2 \tag{2.28}$$

Another second moment bound has been given by Tjur (1990). It can be written as

$$U_2 = \bar{e} - \frac{(1-\bar{e})S_2}{(v-1)(1-\bar{e}) - S_2} \tag{2.29}$$

If the design is efficiency-balanced then $S_2 = 0$ and $U_1 = U_2 = \bar{e}$.

The second moment bounds U_1 and U_2 require knowledge of S_2. However, for equal replicate and equal block size binary designs it is possible to find a lower bound for S_2 which then leads to an upper bound for E based solely on the design parameters; note that both U_1 and U_2 are decreasing functions of S_2. It can be shown that

$$r^2 k^2 S_2 = \sum_{i \neq j} \sum (\lambda_{ij} - \bar{\lambda})^2$$

where λ_{ij} is the (ij)th element of $\mathbf{NN}'$ and $\bar{\lambda} = r(k-1)/(v-1)$ is the mean of the λ_{ij} over all $i \neq j$. Hence, a lower bound for S_2 for binary designs occurs when the off-diagonal elements of $\mathbf{NN}'$ differ by at most one. This gives

$$S_2 \geq S_{2L} = v(v-1)\alpha(1-\alpha)/(rk)^2 \tag{2.30}$$

where α is the fractional part of $\bar{\lambda}$. Upper bounds for E are now given by U_1 and U_2 with S_2 replaced by S_{2L}.

Essentially U_1 and U_2 are making use of the information available on the number of circuits of length 2, c_2, in the treatment concurrence graph of the design; it can be shown that S_2 is a monotonic function of c_2. It follows in the light of the discussion in Section 2.5 that tighter bounds might be expected if information on the number of circuits of lengths $3, 4, \ldots$ was used. Jarrett (1983) gives a third moment bound which uses the number of triangles c_3. Let S_2 be defined as in (2.28) and let $S_3 = \sum(e_i - \bar{e})^3$, then the bound is

$$U_3 = \bar{e} - \frac{S_2^2}{(v-1)(S_3 + \bar{e}S_2)} \qquad (2.31)$$

This bound requires knowledge of both S_2 and S_3. Jarrett (1983) substitutes bounds for S_2 and S_3 into U_3 to give an upper bound in terms of the design parameters for 2-concurrence designs, i.e. designs for which each of the off-diagonal elements of $\mathbf{NN'}$ takes one of two possible values. Of particular interest is the case where the off-diagonal elements differ by one. In the absence of efficiency-balanced designs the class of such designs are, of course, (M,S)-optimal and have been conjectured to include the A-optimal designs; see Section 2.5. With this constraint, a bound for S_3 is given by

$$S_3 \geq S_{3J} = \alpha v(v-1)z/(rk)^3 \qquad (2.32)$$

where

$$z = \begin{cases} \alpha\{(v+1)\alpha - 3\} & \text{if } \alpha < v/\{2(v-1)\} \\ (1-\alpha)\{v - (v+1)\alpha\} & \text{if } \alpha \geq v/\{2(v-1)\} \end{cases}$$

Substituting (2.30) and (2.32) into (2.31) gives a bound U_4, which is conjectured to be an upper bound for E. Jarrett (1983) showed that this bound is a tighter upper bound than U_1 and a bound given earlier by Williams and Patterson (1977), which also used the constraint that the off-diagonal elements of $\mathbf{NN'}$ do not differ by more than one.

For 2-concurrence designs with concurrences differing by one, if the integer part of $\bar{\lambda}$ is equal to zero then Paterson (1983) gives the following bound for S_3

$$S_3 \geq S_{3P} = \alpha v(v-1)[(v+1)\alpha^2 - 3\alpha - k + 2]/(rk)^3 \qquad (2.33)$$

Again substituting (2.30) and (2.33) into (2.31) gives another third moment bound U_5, which is conjectured to be an upper bound for E.

It is possible for the second order bound U_2 to be tighter than the third order bounds U_4 and U_5. The reason for this is that S_2 and S_3 cannot be evaluated exactly, so that it is necessary to substitute bounds for them.

Example 2.3

Suppose an upper bound on E is required for designs with $v = 27$, $k = 3$ and $r = 6$. The arithmetic mean bound of (2.26) is $\bar{e} = 0.6923$. Use of (2.30) in (2.27) and (2.29) gives $U_1 = 0.6773$ and $U_2 = 0.6701$ respectively, while use of (2.30) and (2.32) and of (2.30) and (2.33) in (2.31) gives $U_4 = 0.6745$ and $U_5 = 0.6728$ respectively. In this particular case, therefore, the second moment bound U_2 gives the tightest bound.

The upper bounds given above are best used when $v \leq b$. When $v > b$ tighter bounds can be obtained if use is made of the relationships given in Section 2.7 between a design and its dual. Instead of using the above bounds directly, an appropriate upper bound is first obtained for the dual design using either U_2, U_4 or U_5. The chosen bound is then used in (2.22) in place of E_d.

Example 2.4

Direct use of the above bounds for designs with $v = 12$, $k = 4$, $r = 2$ and $b = 6$ gives $U_2 = 0.7557$ as the tightest bound. For the dual design with $v_d = 6$, $k_d = 2$, $r_d = 4$ and $b_d = 12$ the best bound is $U_4 = 0.5769$. Substituting this value for E_d in (2.22) leads to the improved bound 0.7500 for the $v = 12$ designs.

For non-binary designs, having $v < k$, trace($\mathbf{A}^*$) may vary from one design to another. Williams, Patterson and John (1976) show that the arithmetic mean of the canonical efficiency factors is maximized when there are either k_0 or $k_0 + 1$ replications of each treatment in each block, where k_0 is the integer part of k/v. This leads to the bound given by

$$U_0 = 1 - \frac{k'(v - k')}{k^2(v - 1)} \qquad (2.34)$$

where $k' = k$ modulo v. For binary designs $U_0 = \bar{e}$ but for non-binary designs $U_0 \leq \bar{e}$.

Further, Jarrett (1989) shows that the information matrix $\mathbf{A}$ of the non-binary design can be written as

$$\mathbf{A} = r(1 - \phi^2)(\mathbf{I} - \mathbf{K}) + (r/r')\phi^2 \mathbf{A}_b$$

where $\mathbf{A}_b$ is the information matrix of the binary design derived by deleting k_0 replications of each treatment from each block, $r' = bk'/v$ is the number of times each treatment is replicated in the derived binary design, and where $\phi = k'/k$. It then follows that

$$e_i = 1 - \phi^2(1 - e'_i)$$

where e_i and e'_i are canonical efficiency factors of the non-binary and derived binary design respectively. Hence,

$$S_2 = \phi^4 S'_2, \qquad S_3 = \phi^6 S'_3 \qquad (2.35)$$

where S'_2 and S'_3 are corrected second and third moments respectively of the derived binary design.

Thus, (2.30), (2.32) and (2.33) can be used to obtain bounds for the second and third moments of the derived binary design, and these can be used in (2.35) to obtain bounds for the non-binary design. Finally, with U_0 replacing $\bar{e}$, they can be used in U_1, U_2, U_4 and U_5 to produce upper bounds for the efficiency factor E of non-binary designs.

Example 2.5
For non-binary designs for $v = 9$, $k = 12$ and $r = 8$, $U_0 = 0.9844$. The bounds for the second and third moments, obtained from the derived binary design with $k' = 3$, are

$$S_{2L} = 0.5\phi^4 \qquad S_{3J} = 0.1667\phi^6 \qquad S_{3P} = 0.0$$

where $\phi = 0.25$. Thus, $U_1 = U_2 = U_4 = U_5 = 0.9841$, to four decimal places.

CHAPTER 3

Cyclic designs

3.1 Introduction

Some methods for constructing balanced incomplete block designs
were given in Section 1.2.1. Many of these designs can also be
obtained very simply by a method of cyclic substitution. For
example, a balanced incomplete block design for seven treatments
in blocks of three is given by the seven blocks

$$
\begin{array}{ccccccc}
0 & 1 & 2 & 3 & 4 & 5 & 6 \\
1 & 2 & 3 & 4 & 5 & 6 & 0 \\
3 & 4 & 5 & 6 & 0 & 1 & 2
\end{array}
$$

Each treatment is replicated three times and every pair of
treatments occurs together in a single block. The cyclic method
of construction is such that a block is obtained by adding one to
each element in the previous block and reducing modulo 7 when
necessary. The full design can, therefore, be generated from one of
the blocks, which is called the *initial block*. For convenience, the
initial block will be taken to be the block of lowest numerical value.
Hence, the above design is constructed by a cyclic development of
the initial block (0 1 3).

However, balanced incomplete block designs only exist for a
limited number of combinations of the parameters v, k and r. If
a design with six treatments in six blocks of three is required, for
example, no balanced incomplete block design can be constructed.
However, the method of cyclic substitution can still be used to
provide a design of the required size. One such design is given by
cyclic development of the initial block (0 1 3) to give

$$
\begin{array}{cccccc}
0 & 1 & 2 & 3 & 4 & 5 \\
1 & 2 & 3 & 4 & 5 & 0 \\
3 & 4 & 5 & 0 & 1 & 2
\end{array}
$$

It can be seen that some pairs of treatments, such as 0 and 1, occur together in a single block, while other pairs of treatments, such as 0 and 3, occur together in two blocks. In fact, for this design, the six treatments fall into three groups (0 3), (1 4) and (2 5), with treatments from the same group occurring together in two blocks and those from different groups in a single block. Other cyclic designs of the same size can be generated by taking different initial blocks. The use of the efficiency factors and optimality criteria of the previous chapter can be used to choose between them.

The method of cyclic substitution can be used to construct a flexible class of incomplete block designs. Many balanced incomplete block designs are cyclic, but many are not. Cyclic designs, together with variants based on the cyclical method of construction, will be considered in detail in this chapter.

3.2 Cyclic designs

Cyclic designs are incomplete block designs consisting in the simplest case of a set of blocks obtained by cyclic development of an initial block. More generally, they consist of combinations of such sets and will be said to be of size (v, k, r), where v is the number of treatments, k the block size and r the number of replications. Cyclic designs exist for any combination of parameters that satisfy (1.1). When $r < k$, however, cyclic designs can be inefficient or even disconnected. For such cases it will usually be better to obtain the design from the dual cyclic design or to use some other generalized cyclic design; see Section 3.6.

For given v and k the $\begin{pmatrix} v \\ k \end{pmatrix}$ distinct blocks can be set out in a number of cyclic sets. For example, for $v = 7$ and $k = 3$ the 35 distinct blocks can be set out in five cyclic sets each of $b = 7$ blocks as follows:

```
0123456   0123456   0123456   0123456   0123456
1234560   1234560   1234560   1234560   2345601
2345601   3456012   4560123   5601234   4560123
```

The five sets are, therefore, generated from initial blocks (0 1 2), (0 1 3), (0 1 4), (0 1 5) and (0 2 4) respectively. These sets can be used singly to give cyclic designs with three replications, or in combination to give cyclic designs with $6, 9, \ldots$ replications. That is, cyclic designs of size $(7, 3, r)$ can be constructed for any r which is a multiple of three.

The five sets above are *full sets* consisting of $b = v$ blocks. If v and k are not relatively prime then *partial sets* consisting of v/d blocks arise, where d is any common division of v and k. For example, with $v = 8$ and $k = 4$ the 70 distinct blocks can be set out in eight full sets of eight blocks, one half set of four blocks given by

$$
\begin{array}{cccc}
0 & 1 & 2 & 3 \\
1 & 2 & 3 & 4 \\
4 & 5 & 6 & 7 \\
5 & 6 & 7 & 0
\end{array}
$$

and one quarter set of two blocks given by

$$
\begin{array}{cc}
0 & 1 \\
2 & 3 \\
4 & 5 \\
6 & 7
\end{array}
$$

Hence, in this case, the full and partial sets can be used to give cyclic designs for any value of $r \geq 1$, although the design for $r = 1$ is clearly disconnected.

In general, a partial set is constructed by taking as the initial block a subgroup of the treatments of order d together with $(k/d) - 1$ of its cosets. With $p = v/d$, the treatments

$$0, p, 2p, \ldots, (d - 1)p$$

constitute a subgroup of the v treatment labels closed to addition modulo v. The treatments

$$j, j + p, j + 2p, \ldots, j + (d - 1)p \quad (j = 1, 2, \ldots, p - 1)$$

form the cosets of this subgroup. For example, with $v = 8$ and $k = 4$, the subgroup of size $d = 2$ is 0, 4 with cosets 1, 5; 2, 6 and 3, 7. Note that if the initial block contains the subgroup 0, 4 and coset 2, 6 the partial set will have two, not four, distinct blocks as 0, 2, 4, 6 is itself a subgroup of order 4.

Full and partial sets can be used singly or in combination, thereby giving a flexible class of designs. A cyclic design of size $(v, k, r = ik)$ exists for all positive integers v, k and i. If v and k have a common divisor d then a partial set of size $(v, k, r = k/d)$ exists corresponding to each d. These partial sets may also be combined with the full sets or with other partial sets to form further designs.

Apart from their flexibility, cyclic designs have other advantages which make them attractive. No plan of the experimental layout is needed since the initial blocks are sufficient. This also leads to ease in experimentation. Once treatment labels have been randomly assigned, constant reference to an experimental plan becomes unnecessary. In view of their method of generation cyclic designs of size $(v, k, r = ik)$ provide automatic elimination of heterogeneity in two directions. In the sets for $v = 7$ and $k = 3$, for instance, it can be seen that each treatment appears once in each position within a block. Consequently, position effects can readily be eliminated in the analysis if desired. This point is considered more fully in Chapter 5 on row–column designs. Finally, since the information matrix of a cyclic design is circulant an explicit expression for the canonical efficiency factors can be obtained.

For given v, k and r the most efficient cyclic designs can be determined by simply evaluating and comparing the efficiency factors for each design in the class. The amount of computation necessary can be drastically reduced if use is made of the fact that many cyclic sets and designs are equivalent, or *isomorphic*, to each other, i.e. can be derived from each other by a relabelling or permutation of the treatments.

As an example, consider again the five sets of size (7,3,3) given above. Suppose that treatment i in the set with initial block (0 1 2) is replaced by treatment $3i$, modulo v, for all values of i. The resulting design will be the cyclic set with initial block (0 1 4). Set (0 1 2) is, therefore, equivalent to set (0 1 4). Repeated application of this permutation will establish that the five sets fall into two equivalence groups, with (0 1 2), (0 1 4) and (0 2 4) in one and (0 1 3) and (0 1 5) in the other.

Full details of this permutation theory is given in the catalogue of cyclic designs by John, Wolock and David (1972). Such a theory is an invaluable aid in the search for good designs, since it leads to a considerable reduction in the number of designs that have to be considered.

3.3 Efficiency factors of cyclic designs

A feature arising from the method of constructing cyclic sets is that the k successive differences between treatment labels in any block are invariant throughout the set. This means that, for instance, if treatments 0 and 1 occur together in, say, λ_1 blocks then so also will treatments 1 and 2, 2 and 3, and so on.

Let treatments 0 and i occur together in λ_i blocks ($i = 1, 2, \ldots, v - 1$) and let $\lambda_0 = r$. Then the concurrence matrix $\mathbf{NN}'$ will take the form

$$\mathbf{NN'} = \begin{pmatrix} \lambda_0 & \lambda_1 & \lambda_2 & \cdots & \lambda_{v-1} \\ \lambda_{v-1} & \lambda_0 & \lambda_1 & \cdots & \lambda_{v-2} \\ \lambda_{v-2} & \lambda_{v-1} & \lambda_0 & \cdots & \lambda_{v-3} \\ \vdots & \vdots & \vdots & \ddots & \vdots \\ \lambda_1 & \lambda_2 & \lambda_3 & \cdots & \lambda_0 \end{pmatrix}$$

where $\lambda_i = \lambda_{v-i}$ $(i > 0)$ since $\mathbf{NN'}$ is symmetric. This circulant matrix can be specified by the elements in the first row, since the other rows are obtained from the first row by a cyclical rotation.

Let $\mathbf{\Gamma}_h$ be the $v \times v$ circulant matrix whose first row has 1 in the $(h + 1)$th column and zero elsewhere. Then $\mathbf{NN'}$ can be written more concisely as

$$\mathbf{NN'} = \sum_{h=0}^{v-1} \lambda_h \mathbf{\Gamma}_h$$

Thus, the information matrix $\mathbf{A}$ given in (1.44) can be written as

$$\mathbf{A} = \sum_{h=0}^{v-1} a_h \mathbf{\Gamma}_h \qquad (3.1)$$

where

$$a_0 = r(k-1)/k, \qquad a_h = -\lambda_h/k \qquad (h = 1, 2, \ldots, v-1)$$

Since $\mathbf{A}$ is a symmetric matrix then, using (A.33) in the Appendix, the canonical efficiency factors of a connected cyclic design are given by

$$e_u = (k-1)/k - (1/rk) \sum_{h=1}^{v-1} \lambda_h \cos(2\pi hu/v) \qquad (u = 1, 2, ..., v-1) \qquad (3.2)$$

The Moore–Penrose generalized inverse of $\mathbf{A}$ given in (1.19) can be written as

$$\mathbf{A}^+ = \sum_{h=0}^{v-1} \psi_h \mathbf{\Gamma}_h$$

where

$$\psi_h = (1/rv) \sum_{u=1}^{v-1} e_u^{-1} \cos(2\pi hu/v) \qquad (h = 0, 1, ..., v-1) \qquad (3.3)$$

see (A.36) and (A.37). Hence, the efficiency factor e_{ij} for the

Table 3.1. *Efficiency factors for cyclic designs with $v = 11$, $k = r = 4$*

Initial block	E	e_{min}	e_{max}
(0 1 2 3)	0.702	0.348	0.991
(0 1 2 4)	0.788	0.535	0.948
(0 1 2 5)	0.817	0.707	0.932
(0 1 3 4)	0.794	0.605	0.997

pairwise comparison $\tau_i - \tau_j$ ($i < j$) is, using (2.1), (3.3) and the fact that $\mathbf{A}^+$ is circulant,

$$e_{ij} = \{r(\psi_0 - \psi_{j-i})\}^{-1} \tag{3.4}$$

Note also that

$$E = \frac{v-1}{rv\psi_0} \tag{3.5}$$

where E is the harmonic mean of the canonical efficiency factors or of the pairwise efficiency factors.

The efficiency factors given in (3.2) and (3.4) can be easily calculated. It is unnecessary with cyclic designs to use general, and time-consuming, eigenvalue and inverse subroutines.

Example 3.1
Consider cyclic designs of size (11,4,4). The 330 distinct blocks can be set out in 30 cyclic sets. Under a permutation of the treatment labels each of these sets can be shown to be equivalent to one of four sets. The four designs together with average efficiency factor E and the smallest (e_{min}) and largest (e_{max}) canonical efficiency factors are shown in Table 3.1. There are large differences between the efficiency factors of the designs. The design with initial block (0 1 2 5) is clearly the best design, on the basis of the optimality criteria given in Section 2.4.

3.4 Construction of efficient designs

It has already been seen in (3.2) that the off-diagonal elements of the concurrence matrix $\mathbf{NN}'$ are required in order to calculate the canonical efficiency factors. Further, according to the criteria in Section 2.5 for choosing efficient designs, the set of cyclic designs with high efficiency factors are those which minimize the range of these off-diagonal elements. The search for good designs is facilitated by the fact that the values of the λ_i $(i > 0)$ can be readily obtained from the initial blocks of the design by a method of differencing.

For the full cyclic set with initial block $(0 \; x_1 \; x_2 \ldots x_{k-1})$ all possible differences between pairs of treatments are obtained to give the difference set

$$[x_1, x_2, \ldots, x_{k-1}, x_2 - x_1, \ldots, x_{k-1} - x_1, \ldots, x_{k-1} - x_{k-2}]$$

Any difference i greater than m is replaced by $v - i$, where $m = (v - 1)/2$ for v odd and $m = v/2$ for v even. If a_i represents the number of differences equal to i then this difference set can be represented more briefly by

$$[1^{a_1}, 2^{a_2}, \ldots, m^{a_m}] \tag{3.6}$$

Then $\lambda_i = \lambda_{v-i} = a_i$ for all i except that $\lambda_m = 2a_m$ when v is even. For example, for $v = 7$ and $k = 4$ the initial block $(0 \; 1 \; 3 \; 5)$ gives the paired difference set $[1 \; 3 \; 2 \; 2 \; 3 \; 2]$ or $[1^1, 2^3, 3^2]$. Hence, $\lambda_1 = \lambda_6 = 1$, $\lambda_2 = \lambda_5 = 3$ and $\lambda_3 = \lambda_4 = 2$. For $v = 8$ and $k = 4$ the initial block $(0 \; 1 \; 4 \; 6)$ gives $[1 \; 4 \; 2 \; 3 \; 3 \; 2]$ or $[1^1, 2^2, 3^2, 4^1]$ so that $\lambda_1 = \lambda_7 = 1$, $\lambda_2 = \lambda_6 = 2$, $\lambda_3 = \lambda_5 = 2$ and $\lambda_4 = 2$.

For partial sets with v/d blocks, where d is a common divisor of v and k, the same differencing procedure is used except that the λ_i obtained from (3.6) have to be further divided by d. For example, for $v = 8$ and $k = 4$ a partial set with four blocks $(d = 2)$ is obtained from the initial block $(0 \; 1 \; 4 \; 5)$. Since this gives $[1 \; 4 \; 3 \; 3 \; 4 \; 1]$ or $[1^2, 2^0, 3^2, 4^2]$ then $\lambda_1 = \lambda_7 = 1$, $\lambda_2 = \lambda_6 = 0$, $\lambda_3 = \lambda_5 = 1$ and $\lambda_4 = 2$. For designs with more than one initial block the concurrences are obtained by adding together the λ_i values from the separate initial blocks. A design of size $(8,4,6)$ with initial blocks $(0 \; 1 \; 4 \; 5)$ and $(0 \; 1 \; 4 \; 6)$ has, therefore, $\lambda_1 = \lambda_2 = \lambda_6 = \lambda_7 = 2$, $\lambda_3 = \lambda_5 = 3$ and $\lambda_4 = 4$.

This method of differencing allows good cyclic designs to be constructed, in particular (M,S)-optimal designs are readily obtained. A search for designs can, for example, be restricted to the class of regular graph designs, which have treatments occurring together in either λ_1 or λ_2 blocks, where λ_1 and λ_2 differ by at most one. For instance, the design of size $(12,4,6)$ with initial blocks $(0 \; 1 \; 6 \; 7)$ and $(0 \; 2 \; 4 \; 11)$ is a regular graph design, since it can be verified that pairs of treatments occur together in either one or two blocks. For another example, consider constructing a regular graph cyclic design of size $(12,4,8)$, made up of two full sets. The average value of the off-diagonal elements in the concurrence matrix $\mathbf{NN}'$ is $\lambda = r(k-1)/(v-1) = 24/11$. Hence, treatment 0 must occur with n_1 treatments in $\lambda_1 = 2$ blocks and with the remaining n_2 treatments in $\lambda_2 = 3$ blocks, where $n_1 + n_2 = v - 1$ and $n_1\lambda_1 + n_2\lambda_2 = r(k-1)$. Thus $n_1 = 9$ and $n_2 = 2$. Let the first initial block be (0 1 3 7), which has difference set $[1^1, 2^1, 3^1, 4^1, 5^1, 6^2]$. There are a number of possible choices for the initial block of the second set, e.g. (0 1 2 5), (0 2 4 5) and (0 1 3 8), all of which produce the desired regular graph design. Other designs will also be obtained by changing the first initial block. The average efficiency factor E, given in (3.5), can be used to choose among the designs in this regular graph class. It should be noted, however, that cyclic regular graph designs do not exist for all parameter combinations, especially if partial sets are involved; for instance, no such design of size (8,4,6) exists.

Cyclic designs with $r < k$ can also be relatively inefficient. This is because the initial block of a partial set of $p = v/d$ blocks, where d is a common divisor of v and k, must consist of the subgroup $0, p, 2p, \ldots, (d-1)p$ and a number of its cosets, as explained in Section 3.2. Thus, from the difference set (3.6), it follows that $\lambda_p = \lambda_{2p} = \ldots = \lambda_{(d-1)p} = k/d$ which may be considerably different from other λ_i values. When $r < k$ it is usually preferable, therefore, to use the dual of a cyclic design for $r > k$. Although these dual designs may not be cyclic, many can be obtained by cyclical methods of construction as is shown in Section 3.6.

A number of catalogues of efficient cyclic designs are available. John, Wolock and David (1972) provide extensive tables of designs for $v \leq 30$ and $r \leq 10$. Lamacraft and Hall (1982) give a full listing of designs in the range $10 \leq v \leq 60$, $3 \leq r = k \leq 10$. John (1981) provides two small compact tables of efficient cyclic designs which can be used whenever $6 \leq v \leq 30$, $r \leq 10$, and r is a multiple of k.

The availability of extremely fast computer algorithms to generate optimal, or near-optimal, cyclic designs provides an attractive alternative to these catalogues of designs. John, Whitaker and Triggs (1993) formulate the problem of constructing efficient cyclic designs as a non-linear integer programme, and use a branch and bound algorithm to generate the designs. Designs involving more than one cyclic set can be built up sequentially if necessary, and designs for $r < k$ can be obtained as the duals of cyclic designs with $r > k$. More recently a faster algorithm has been described by Nguyen (1994), which automatically uses the dual approach when $r < k$.

3.5 n-Cyclic designs

The cyclic designs considered in the previous sections have been obtained by a cyclic development of one or more initial blocks. Two important features of a cyclic design are that each treatment label is represented by a single number and that the elements of a block are obtained by adding one to the elements in the previous block, and reducing modulo v when necessary. Two generalizations will now be considered. In the first, each treatment label is represented by a set of numbers and the cyclic development is carried out on each number in turn. The resulting designs will be called n-cyclic designs and will be considered in this section; these designs have also been called GC/n designs. In the second generalization a value greater than one is added to the elements of a block to give the next block. These designs will be called *generalized cyclic* designs and will be considered in Section 3.6.

Two of the balanced incomplete block designs listed in the tables of Fisher and Yates (1963) are given by dicyclic (i.e. 2-cyclic) solutions. One of these designs has 16 treatments in 16 blocks of six. The 16 treatments are represented by four letters in combination with four suffixes. Four blocks are first obtained by cyclic development of the letters in the initial block $(a_1 \ a_2 \ a_3 \ b_1 \ c_4 \ d_1)$. The full dicyclic design is then given by a cyclic development of the suffixes in each of these four blocks.

Alternatively, each treatment can be represented by a pair of digits $a_1 a_2$, say, where $a_i = 0, 1, 2, 3 \ (i = 1, 2)$. The balanced incomplete block design is then obtained from the initial block (00 01 02 10 23 30) by a cyclic development of each digit in turn. The 16 blocks are

00 01 02 03	10 11 12 13	20 21 22 23	30 31 32 33
01 02 03 00	11 12 13 10	21 22 23 20	31 32 33 30
02 03 00 01	12 13 10 11	22 23 20 21	32 33 30 31
10 11 12 13	20 21 22 23	30 31 32 33	00 01 02 03
23 20 21 22	33 30 31 32	03 00 01 02	13 10 11 12
30 31 32 33	00 01 02 03	10 11 12 13	20 21 22 23

Note that the 16 blocks fall into four groups of four blocks each. The first blocks in the groups are obtained from the initial block by cyclic development of the first digit. The blocks within each group are then obtained from the first block of the group by cyclic development of the second digit.

Each treatment label in a dicyclic design is made up of two digits, and as such is said to be a two-factor design. More generally, an

n-cyclic design will have n factors and $v = m_1 m_2 \ldots m_n$ treatments, where the ith factor is said to be at m_i levels ($i = 1, 2, \ldots, n$). A treatment is represented by an n-tuple $a_1 a_2 \ldots a_n$ where $a_i = 0, 1, \ldots, m_i - 1$ ($i = 1, 2, \ldots, n$). Treatments can be written in order of magnitude and identified with the numbers $1, 2, \ldots, v$. For example, with $n = 3$, $m_1 = 2$, $m_2 = 2$ and $m_3 = 3$ the $v = 12$ treatments are

000 001 002   010 011 012   100 101 102   110 111 112

so that, for instance, the third treatment is 002 and the eighth 101.

An n-cyclic set is generated from an initial block consisting of k treatments. The jth block of the set is given by adding the jth treatment to each treatment in the initial block, where addition is defined as

$$a_1 a_2 \ldots a_n + b_1 b_2 \ldots b_n = c_1 c_2 \ldots c_n \qquad (3.7)$$

where $c_i = a_i + b_i$ modulo m_i ($i = 1, 2, \ldots, n$). This method of generating a set is equivalent to a cyclical development of the initial block where each digit in the n-tuple is cycled in turn.

Example 3.2
As a further example, consider the 3-cyclic set for $v = 12$, $n = 3$, $m_1 = 2$, $m_2 = 2$, $m_3 = 3$ and $k = 4$ obtained from the initial block (000 011 101 112). The full set of 12 blocks is

```
000  001  002  010  011  012  100  101  102  110  111  112
011  012  010  001  002  000  111  112  110  101  102  100
101  102  100  111  112  110  001  002  000  011  012  010
112  110  111  102  100  101  012  010  011  002  000  001
```

The two designs given above are examples of full sets with $b = v$ blocks. Partial sets can also be constructed using a method similar to that used in Section 3.2 to obtain partial cyclic sets. The v treatment labels form a group G closed to addition defined by (3.7). Let S be a subgroup of G of order d, where d is a common factor of v and k. The initial block of a partial set of v/d blocks is then given by S together with $(k/d) - 1$ of its cosets. For instance, for $v = 12$, $n = 3$, $m_1 = m_2 = 2$, $m_3 = 3$ and $k = 4$ a subgroup of order 2 is given by 000 110 while a subgroup of order 4 is given by 000 010 100 110. Cosets are obtained by adding further treatments to every element in the subgroup. Hence, for example, initial blocks for sets with six and three blocks are (000 001 110 111) and (000 010 100 110) respectively.

By taking partial or full sets singly or in combination a large and flexible class of n-cyclic designs is available. Two of the balanced incomplete block designs listed in the tables of Fisher and Yates (1963) are 2-cyclic designs. One is given above. The other has 25 treatments in 50 blocks of four and is made up of two 2-cyclic sets with initial blocks (00 01 10 44) and (00 02 20 33), where $m_1 = m_2 = 5$.

n-Cyclic designs have been shown to be particularly useful in the construction of designs for factorial experiments, as will be shown in Chapters 8 and 9.

In the analysis of n-cyclic designs, the concurrence matrix $\mathbf{NN}'$, the information matrix $\mathbf{A}$ of (1.14) and its generalized inverse matrix $\mathbf{\Omega}$ given by (1.19) or (1.45) will all be block circulant matrices. The notation used in Section 3.3 for circulant matrices can be extended to block circulants; see also Section A.8 in the Appendix.

To fix ideas, consider the concurrence matrix $\mathbf{NN}'$ for the design for $v = b = 12$ and $r = k = 4$ given in Example 3.2, namely

$$\mathbf{NN}' = \left(\begin{array}{ccc|ccc|ccc|ccc}
4 & 0 & 0 & 0 & 2 & 2 & 0 & 2 & 2 & 2 & 1 & 1 \\
0 & 4 & 0 & 2 & 0 & 2 & 2 & 0 & 2 & 1 & 2 & 1 \\
0 & 0 & 4 & 2 & 2 & 0 & 2 & 2 & 0 & 1 & 1 & 2 \\
\hline
0 & 2 & 2 & 4 & 0 & 0 & 2 & 1 & 1 & 0 & 2 & 2 \\
2 & 0 & 2 & 0 & 4 & 0 & 1 & 2 & 1 & 2 & 0 & 2 \\
2 & 2 & 0 & 0 & 0 & 4 & 1 & 1 & 2 & 2 & 2 & 0 \\
\hline
0 & 2 & 2 & 2 & 1 & 1 & 4 & 0 & 0 & 0 & 2 & 2 \\
2 & 0 & 2 & 1 & 2 & 1 & 0 & 4 & 0 & 2 & 0 & 2 \\
2 & 2 & 0 & 1 & 1 & 2 & 0 & 0 & 4 & 2 & 2 & 0 \\
\hline
2 & 1 & 1 & 0 & 2 & 2 & 0 & 2 & 2 & 4 & 0 & 0 \\
1 & 2 & 1 & 2 & 0 & 2 & 2 & 0 & 2 & 0 & 4 & 0 \\
1 & 1 & 2 & 2 & 2 & 0 & 2 & 2 & 0 & 0 & 0 & 4
\end{array}\right)$$

The matrix can be partitioned into four 6×6 matrices with each of these matrices partitioned into four 3×3 matrices, as indicated. This shows the block circulant structure of the matrix. Thus

$$\mathbf{NN}' = \left(\begin{array}{cc} \mathbf{B}_0 & \mathbf{B}_1 \\ \mathbf{B}_1 & \mathbf{B}_0 \end{array}\right)$$

where $\mathbf{B}_h$ is a 6×6 matrix ($h = 0, 1$). Hence

$$\mathbf{NN}' = \begin{pmatrix} 1 & 0 \\ 0 & 1 \end{pmatrix} \otimes \mathbf{B}_0 + \begin{pmatrix} 0 & 1 \\ 1 & 0 \end{pmatrix} \otimes \mathbf{B}_1 = \sum_{h_1=0}^{1} (\mathbf{\Gamma}_{h_1} \otimes \mathbf{B}_{h_1})$$

where $\mathbf{\Gamma}_{h_i}$ is the $m_i \times m_i$ circulant matrix whose first row has 1 in the $(h + 1)$th column and zero elsewhere, and where $\otimes$ denotes the Kronecker product (see Section A.6 of the Appendix). Further, each $\mathbf{B}_{h_1}$ matrix can be written as

$$\mathbf{B}_{h_1} = \begin{pmatrix} \mathbf{B}_{h_10} & \mathbf{B}_{h_11} \\ \mathbf{B}_{h_11} & \mathbf{B}_{h_10} \end{pmatrix} = \sum_{h_2=0}^{1} (\mathbf{\Gamma}_{h_2} \otimes \mathbf{B}_{h_1 h_2})$$

where $\mathbf{B}_{h_1 h_2}$ is a 3×3 matrix. Finally, each $\mathbf{B}_{h_1 h_2}$ matrix is circulant so that

$$\mathbf{B}_{h_1 h_2} = \sum_{h_3=0}^{2} \lambda_{h_1 h_2 h_3} \mathbf{\Gamma}_{h_3}$$

where the $\lambda_{h_1 h_2 h_3}$ are elements of the first row of $\mathbf{B}_{h_1 h_2}$. Therefore, the concurrence matrix for the above 3-cyclic design can be written as

$$\mathbf{NN}' = \sum_{h_1=0}^{1} \mathbf{\Gamma}_{h_1} \otimes \left(\sum_{h_2=0}^{1} \mathbf{\Gamma}_{h_2} \otimes \sum_{h_3=0}^{2} \lambda_{h_1 h_2 h_3} \mathbf{\Gamma}_{h_3} \right)$$

i.e.

$$\mathbf{NN}' = \sum_{h_1=0}^{1} \sum_{h_2=0}^{1} \sum_{h_3=0}^{2} \lambda_{h_1 h_2 h_3} (\mathbf{\Gamma}_{h_1} \otimes \mathbf{\Gamma}_{h_2} \otimes \mathbf{\Gamma}_{h_3})$$

where

$$r = \lambda_{000} = 4, \lambda_{001} = \lambda_{002} = 0, \lambda_{010} = 0, \lambda_{011} = \lambda_{012} = 2$$
$$\lambda_{100} = 0, \lambda_{101} = \lambda_{102} = 2, \lambda_{110} = 2, \lambda_{111} = \lambda_{112} = 1$$

More generally, the block circulant concurrence matrix of an n-cyclic design can be written as

$$\mathbf{NN}' = \sum_{h_1=0}^{m_1-1} \sum_{h_2=0}^{m_2-1} \cdots \sum_{h_n=0}^{m_n-1} \lambda_{h_1 h_2 \ldots h_n} (\mathbf{\Gamma}_{h_1} \otimes \mathbf{\Gamma}_{h_2} \otimes \ldots \otimes \mathbf{\Gamma}_{h_n}) \quad (3.8)$$

By an extension of the methods used to obtain (3.2), explicit expressions can be obtained for the canonical efficiency factors of an n-cyclic design. They are

$$e_{u_1...u_n} = 1 - (1/rk) \sum_{h_1=0}^{m_1-1} \cdots \sum_{h_n=0}^{m_n-1} \lambda_{h_1...h_n} \cos\left\{\sum_{j=1}^{n}(2\pi u_j h_j/m_j)\right\}$$

(3.9)

for $u_l = 0, 1, \ldots, m_l - 1$; $l = 1, 2, \ldots, n$. Note that if $n = 1$ then (3.9) is equivalent to (3.2).

3.6 Generalized cyclic designs

Generalized cyclic incomplete block designs were first considered by Jarrett and Hall (1978). The $v = mn$ treatments are divided into $m \, (> 1)$ groups of n elements using the residue classes, modulo m. Thus, the ith residue class is

$$S_i = \{i, i+m, \ldots, i+m(n-1)\} \quad (i = 0, 1, ..., m-1) \quad (3.10)$$

The designs are obtained by successive addition of m, modulo v, to the elements of one or more initial blocks. In general, each initial block contributes n blocks to the design, although again partial sets can be constructed. A generalized cyclic design with increment number m will be noted by GCIB$_m$.

Example 3.3
Suppose there are $v = 8$ treatments divided into two groups of four. The residue classes are

$$\begin{aligned} S_0 &= \{0, 2, 4, 6\} \\ S_1 &= \{1, 3, 5, 7\} \end{aligned}$$

Two GCIB$_2$ designs for $k = 4$ and $b = 4$ can be obtained from the initial blocks (0 1 2 4) and (0 1 2 5). They are

0	2	4	6		0	2	4	6
1	3	5	7		1	3	5	7
2	4	6	0		2	4	6	0
4	6	0	2		5	7	1	3

In the first design the treatments from class S_0 are each replicated three times while those from class S_1 are replicated once. In the second design all treatments are replicated twice.

It follows from the method of construction that all treatments within a residue class are equally replicated, although different residue classes may have different replications. Equal replication of all treatments is achieved if the initial blocks contain the same number of treatments from each residue class. For equal block size designs, the concurrence matrix $\mathbf{NN}'$ is completely determined from the first m rows. Hence, $\mathbf{NN}'$ can be partitioned into $m \times m$ submatrices, $\mathbf{B}_h$ $(h = 0, 1, \ldots, n - 1)$ arranged in a circulant manner, thus

$$\mathbf{NN}' = \sum_{h=0}^{n-1} \mathbf{\Gamma}_h \otimes \mathbf{B}_h$$

where $\mathbf{\Gamma}_h$ is a basic circulant matrix of order n as defined in the previous section. For instance, for the equal replicated design for $v = 8$ given in Example 3.3,

$$\mathbf{NN}' = \left(\begin{array}{cc|cc|cc|cc} 2 & 1 & 1 & 1 & 0 & 1 & 1 & 1 \\ 1 & 2 & 1 & 0 & 1 & 2 & 1 & 0 \\ \hline 1 & 1 & 2 & 1 & 1 & 1 & 0 & 1 \\ 1 & 0 & 1 & 2 & 1 & 0 & 1 & 2 \\ \hline 0 & 1 & 1 & 1 & 2 & 1 & 1 & 1 \\ 1 & 2 & 1 & 0 & 1 & 2 & 1 & 0 \\ \hline 1 & 1 & 0 & 1 & 1 & 1 & 2 & 1 \\ 1 & 0 & 1 & 2 & 1 & 0 & 1 & 2 \end{array} \right)$$

so that

$$\mathbf{B}_0 = \left(\begin{array}{cc} 2 & 1 \\ 1 & 2 \end{array} \right), \quad \mathbf{B}_1 = \mathbf{B}_3 = \left(\begin{array}{cc} 1 & 1 \\ 1 & 0 \end{array} \right) \quad \text{and} \quad \mathbf{B}_2 = \left(\begin{array}{cc} 0 & 1 \\ 1 & 2 \end{array} \right)$$

The matrix $\mathbf{A}$ of (1.14) and its generalized inverse matrix $\mathbf{\Omega}$ given by (1.19) or (1.45) will have the same structure. A similar structure also holds for the more general designs with unequal block sizes. It is not possible to write down explicit expressions for the efficiency factors of generalized cyclic designs in the way that was done for cyclic and n-cyclic designs. Jarrett and Hall (1978) show, however, that the canonical efficiency factors can be obtained by calculating the non-zero eigenvalues of certain $m \times m$ complex matrices.

As has already been indicated in Section 3.4, when $r < k$ cyclic designs will often be relatively inefficient in terms of the overall efficiency factor. In such cases it is frequently possible to find more efficient generalized cyclic designs. For example, consider a comparison of the GCIB$_2$ design given in Example 3.3 for $v = 8$

and initial block (0 1 2 5) with the partial cyclic set of the same size with initial block (0 1 4 5). In the $GCIB_2$ design the number of pairs of treatments occurring together in 0, 1 and 2 blocks are 6, 20 and 2 respectively. For the partial cyclic set, the frequencies of these concurrences are 8, 16 and 4 respectively. Using the simple counting rules given in Section 2.5, therefore, the $GCIB_2$ design will be preferable in that it has concurrences which are more nearly equal; in fact, the average efficiency factor is 0.808 for the $GCIB_2$ design compared with 0.778 for the cyclic design. The incidence matrix for this $GCIB_2$ design is

$$
\mathbf{N} = \begin{pmatrix}
1 & 0 & 0 & 1 \\
1 & 0 & 1 & 0 \\
1 & 1 & 0 & 0 \\
0 & 1 & 0 & 1 \\
0 & 1 & 1 & 0 \\
1 & 0 & 1 & 0 \\
0 & 0 & 1 & 1 \\
0 & 1 & 0 & 1
\end{pmatrix}
$$

Note that the odd numbered rows of $\mathbf{N}$ are obtained by cyclic development of the first row and that the even numbered rows from a cyclic development of the second row. Thus the matrix $\mathbf{N}'$ is the incidence matrix of two full cyclic sets with initial blocks given by the first two rows of $\mathbf{N}$. The dual of this $GCIB_2$ design is, therefore, the cyclic design for four treatments with initial blocks (0 3) and (0 2), with the half set (0 2) replicated twice.

In general, the dual of a $GCIB_m$ design for $v = mn$ treatments in n blocks of k is a cyclic design with n treatments in v blocks of $r = k/m$ obtained from m initial blocks. It follows from (2.22), therefore, that efficient cyclic designs can be used to construct efficient generalized cyclic designs. Using the best cyclic design does not, however, necessarily produce the best generalized cyclic design for, as Hall and Jarrett (1981) have shown, better designs may be obtained in some cases by increasing the value of m. For example, consider the construction of a generalized cyclic design for $v = 32$, $r = 2$, $k = 4$ and $b = 16$. With $m = 2$ and $n = 16$, the best cyclic design for 16 treatments in 32 blocks of two has initial blocks (0 1) and (0 6). The dual of this design is the $GCIB_2$ design with initial block (0 1 2 13) and average efficiency factor $E = 0.619$. A more efficient generalized cyclic design of the same size is obtained from the initial blocks (0 1 2 5) and (0 3 11 14) with increment number $m = 4$. The average efficiency factor of this design is $E = 0.624$.

The above result on the duality of cyclic designs is a special case of a more general result given by Hall and Jarrett (1981). They show that the dual of a GCIB_m design with mn treatments and qn blocks is a GCIB_q design with qn treatments and mn blocks. They use this result to produce tables of efficient generalized cyclic designs for $10 \leq v \leq 60$, $r \leq k$ and for $(r, k) = (2,4)$, $(2,6)$, $(2,8)$, $(2,10)$, $(3,6)$, $(3,9)$, $(4,6)$, $(4,8)$, $(4,10)$, $(5,10)$. For other cases with $r \leq 5$ and $k \leq 10$, r and k have no factors in common and k must necessarily divide v. For these cases they suggest the use of the α-designs of Patterson and Williams (1976a), which are discussed in detail in Chapter 4.

Resolvable block designs

4.1 Introduction

The practical importance of incomplete block designs has been discussed in Section 1.1. The cyclic and generalized cyclic designs of the previous chapter provide a wide range of very efficient incomplete block designs. In many areas of experimentation it is desirable to restrict attention to incomplete block designs where the blocks can be grouped together so that each treatment is replicated exactly once in each group. Such designs are called *resolvable* incomplete block designs.

As an example, consider the balanced incomplete block design given in Example 1.1 for nine treatments in 12 blocks of three, constructed from the complete set of orthogonal 3×3 Latin squares. The design is

```
0  3  6    0  1  2    0  1  2    0  1  2
1  4  7    3  4  5    4  5  3    5  3  4
2  5  8    6  7  8    8  6  7    7  8  6
```

where again the blocks are written as columns. The design has been set out in four groups of blocks such that each treatment is replicated exactly once in each group, i.e. the design is resolvable. Each group of blocks constitutes a *replicate*.

Resolvable designs are important in practice since it is often useful to be able to perform an experiment a replicate at a time. In an industrial experiment, for instance, it may not always be possible to carry out all trials in a single session. Use of resolvable designs permits trials to be carried out in stages, with one or more replicates dealt with at each stage. If the experiment has to be discontinued at any time then all treatments will have occurred equally often. Accuracy will also be increased if the experimental material can be arranged so that replicates are relatively homogeneous. In an agricultural experiment, for example, the land may be divided into a number of large areas corresponding to the replicates and then each area subdivided into blocks. Gains in efficiency can also be achieved by resolvable designs when inter-block information is recovered in the analysis; this point is discussed in Section 7.6.2. Resolvable designs are extensively used in variety trials in the United Kingdom; see Patterson and Silvey (1980).

Some of the balanced incomplete block and cyclic designs already discussed are resolvable, so are a number of other designs in the literature. But in view of the importance of resolvability in practice, it is often desirable to make this property a prerequisite in the construction of incomplete block designs. It was such an approach that led Yates (1936b) to introduce square lattice designs and Harshbarger (1949) rectangular lattice designs. Later α-designs were developed by Patterson and Williams (1976a) to greatly extend the availability of efficient resolvable designs.

The practical value of resolvable designs, in particular α-designs, has been enhanced by the development of

(i) computer software to generate designs for a wide range of parameter values, and

(ii) effective upper bounds by which the average efficiency factor of a design can be assessed against theoretical optima; see Section 4.10.

4.2 Square lattice designs

Square lattice designs are resolvable designs for $v = s^2$ treatments and $k = s$ plots per block. They are constructed as follows. The s^2 treatments are set out in an $s \times s$ array. In the first replicate, rows of the array correspond to blocks, i.e. all treatments in the jth row of the array are placed in the jth block ($j = 1, 2, \ldots, s$). In the second replicate, blocks correspond to columns of the array. For a third replicate, a Latin square is superimposed on the array and all treatments corresponding to the same letter in the Latin square are placed in the same block. For certain values of s, a fourth replicate can be constructed in a similar way by superimposing onto the array a Latin square orthogonal to the first. Further replicates can be obtained by using Latin squares orthogonal to all previous Latin squares, if such squares exist. The method of construction follows that used to construct the balanced incomplete block designs for $v = m^2$ in Section 1.2.1.

Square lattice designs for $r = 2$ and $r = 3$ replicates are known as *simple lattice* and *triple lattice* designs respectively, and can be constructed for all values of s. For $s = 6$ only simple and triple lattice designs can be obtained, as no orthogonal Latin squares exist. Lattice designs based on the complete set of mutually orthogonal Latin squares will have $r = s + 1$ replicates and are known as *balanced lattice* designs; they are, of course, also balanced incomplete block designs. Such designs exist for any $s = p^t$, where p is a prime number.

The design for nine treatments in 12 blocks of three given in the last section is a balanced lattice design for $s = 3$ and $r = 4$. Simple and triple lattice designs, for $s = 3$, are given by taking the first two and three replicates respectively.

The properties of a square lattice design are most easily derived from its dual design, i.e. from the block concurrence matrix $\mathbf{N'N}$ of the square lattice design. Now the (ij)th element of $\mathbf{N'N}$ will be equal to the number of treatments common to both the ith and jth blocks of the square lattice design. For two blocks within the same replicate this element is zero, while for two blocks from different replicates it is 1. Further, the diagonal elements of $\mathbf{N'N}$ are all equal to s. For example, for $s = 3$ the $\mathbf{N'N}$ matrices for square lattice designs with $r = 2, 3, 4$ can be written in the form

$$\begin{pmatrix} 3\mathbf{I} & \mathbf{J} \\ \mathbf{J} & 3\mathbf{I} \end{pmatrix}, \quad \begin{pmatrix} 3\mathbf{I} & \mathbf{J} & \mathbf{J} \\ \mathbf{J} & 3\mathbf{I} & \mathbf{J} \\ \mathbf{J} & \mathbf{J} & 3\mathbf{I} \end{pmatrix}, \quad \begin{pmatrix} 3\mathbf{I} & \mathbf{J} & \mathbf{J} & \mathbf{J} \\ \mathbf{J} & 3\mathbf{I} & \mathbf{J} & \mathbf{J} \\ \mathbf{J} & \mathbf{J} & 3\mathbf{I} & \mathbf{J} \\ \mathbf{J} & \mathbf{J} & \mathbf{J} & 3\mathbf{I} \end{pmatrix}$$

In general, for a square lattice design and provided the blocks are ordered replicate by replicate, the matrix $\mathbf{N'N}$ can be partitioned so that all the diagonal matrices are $k\mathbf{I}_s$ and all off-diagonal matrices are $\mathbf{J}_s$, i.e.

$$\mathbf{N'N} = k\mathbf{I}_r \otimes \mathbf{I}_s + (\mathbf{J}_r - \mathbf{I}_r) \otimes \mathbf{J}_s \tag{4.1}$$

Now the vector $\mathbf{p} = \mathbf{p}_1 \otimes \mathbf{p}_2$, where $\mathbf{p}_1$ and $\mathbf{p}_2$ are $r \times 1$ and $s \times 1$ vectors respectively, is an eigenvector of $\mathbf{N'N}$ if $\mathbf{p}_i$ is a contrast vector or if $\mathbf{p}_i = \mathbf{1}$ $(i = 1, 2)$. A full set of eigenvectors can be obtained by taking both $\mathbf{p}_1$ and $\mathbf{p}_2$ to be contrast vectors or equal to $\mathbf{1}$, or by taking one to be a contrast vector and the other equal to $\mathbf{1}$. The canonical efficiency factors of the dual design can then be shown to be $(r-1)/r$ and 1 with multiplicities $r(s-1)$ and $r-1$ respectively. Hence, using results in Section 2.7, the canonical efficiency factors of a square lattice design are $(r-1)/r$ and 1 with multiplicities $r(s-1)$ and $v-1-r(s-1)$ respectively. For balanced

lattice designs $(r = s + 1)$ all canonical efficiency factors are equal
to $(r - 1)/r$.

It follows, using (2.9), that the average efficiency factor E of a
square lattice design is

$$E = \frac{(s + 1)(r - 1)}{r(s + 2) - (s + 1)} \tag{4.2}$$

For simple, triple and balanced lattice designs E is respectively

$$\frac{s + 1}{s + 3}, \quad \frac{2s + 2}{2s + 5}, \quad \frac{s}{s + 1}$$

Finally, a straightforward generalized inverse Ω of the
information matrix $\mathbf{A}$ of a square lattice design can be obtained
using (2.25). From (2.24) and (4.1),

$$\mathbf{A}_d = s\mathbf{I} - (1/r)\mathbf{I}_r \otimes (s\mathbf{I}_s - \mathbf{J}_s) - (1/r)\mathbf{J}_r \otimes \mathbf{J}_s$$

Following (1.45), a generalized inverse of $\mathbf{A}_d$ is given by Ω_d where

$$\Omega_d^{-1} = \mathbf{A}_d + (1/r)\mathbf{J}_r \otimes \mathbf{J}_s = (1/r)\mathbf{I}_r \otimes \{s(r - 1)\mathbf{I}_s + \mathbf{J}_s\}$$

Hence,

$$\Omega_d = \frac{r}{s(r - 1)}\mathbf{I}_r \otimes \left(\mathbf{I}_s - \frac{1}{rs}\mathbf{J}_s\right)$$

Now, since $\mathbf{N}(\mathbf{I}_r \otimes \mathbf{J}_s) = \mathbf{J}_{v \times rs}$ as each $s \times s$ submatrix in $\mathbf{N}$
contains a single unit element in each row and zeros elsewhere,

$$\mathbf{N}\Omega_d\mathbf{N}' = \frac{r}{s(r - 1)}\left(\mathbf{N}\mathbf{N}' - \frac{1}{s}\mathbf{J}_v\right)$$

From (2.25), and as $\mathbf{AJ} = \mathbf{0}$, it follows that

$$\Omega = \frac{1}{r}\left\{\mathbf{I} + \frac{1}{s(r - 1)}\mathbf{N}\mathbf{N}'\right\} \tag{4.3}$$

The full intra-block analysis for any square lattice design is,
consequently, relatively straightforward. No matrix inversions are
necessary.

4.3 Rectangular lattice designs

Rectangular lattice designs are resolvable designs for $v = s(s - 1)$
treatments and $k = s - 1$ plots per block. The treatments are
initially set out in an $s \times s$ array with the leading diagonal left
blank. The blocks of the first two replicates correspond to the
rows and columns respectively of the array. Further replicates

are obtained by superimposing orthogonal Latin squares onto the array, with the constraint that the leading diagonal of each Latin square contains every treatment in the same order. If there exist $t - 1$ mutually orthogonal Latin squares then $t - 2$ of these squares will satisfy this additional constraint. Blocks in each replicate are obtained, as for square lattice designs, by collecting together all treatments corresponding to the same letter in the particular Latin square. Designs with $r = 2$ and $r = 3$ are known as *simple* and *triple rectangular lattice* designs respectively.

Example 4.1

For $s = 4$, the 12 treatments are set out in a 4×4 array as follows:

$$
\begin{array}{cccc}
\cdot & 0 & 1 & 2 \\
3 & \cdot & 4 & 5 \\
6 & 7 & \cdot & 8 \\
9 & 10 & 11 & \cdot
\end{array}
$$

giving two replicates

$$
\begin{array}{cccc|cccc}
0 & 3 & 6 & 9 & 3 & 0 & 1 & 2 \\
1 & 4 & 7 & 10 & 6 & 7 & 4 & 5 \\
2 & 5 & 8 & 11 & 9 & 10 & 11 & 8
\end{array}
$$

Superimposing the following orthogonal Latin squares onto the array leads to two further replicates

$$
\begin{array}{cccc}
A & C & D & B \\
D & B & A & C \\
B & D & C & A \\
C & A & B & D
\end{array}
\qquad
\begin{array}{cccc}
A & D & B & C \\
C & B & D & A \\
D & A & C & B \\
B & C & A & D
\end{array}
$$

The two replicates are

$$
\begin{array}{cccc|cccc}
4 & 2 & 0 & 1 & 5 & 1 & 2 & 0 \\
8 & 6 & 5 & 3 & 7 & 8 & 3 & 4 \\
10 & 11 & 9 & 7 & 11 & 9 & 10 & 6
\end{array}
$$

Properties of rectangular lattice designs can again be obtained most easily by considering the dual designs. By a suitable ordering of blocks within replicates, the block concurrence matrix $\mathbf{N'N}$ partitions into r^2, $s \times s$ submatrices with all diagonal submatrices equal to $k\mathbf{I}_s$ and all off-diagonal submatrices equal to $\mathbf{J}_s - \mathbf{I}_s$, i.e.

$$\mathbf{N'N} = k\mathbf{I}_r \otimes \mathbf{I}_s + (\mathbf{J}_r - \mathbf{I}_r) \otimes (\mathbf{J}_s - \mathbf{I}_s) \tag{4.4}$$

The canonical efficiency factors of the dual of a rectangular lattice design can be shown to be $(rk - s)/rk$, $s(r - 1)/rk$ and 1 with multiplicities $k(r-1)$, k and $r-1$ respectively. Hence, using results in Section 2.7, the canonical efficiency factors of a rectangular lattice design are, for $r = s$, $(k-1)/k$ and 1 with multiplicities k^2 and $k-1$ respectively and, for $r < s$, $(rk-s)/rk$, $s(r-1)/rk$ and 1 with multiplicities $k(r-1)$, k and $v-1-rk$ respectively. Hence, the average efficiency factor of a rectangular lattice design is, after some algebraic manipulation, given by

$$E = \frac{v-1}{v-1+rk\{kv+(r-s)\mu\}} \tag{4.5}$$

where $\nu = 1/(rk-s)$ and $\mu = \nu/\{s(r-1)\}$.

Williams (1977) shows that a generalized inverse of the information matrix $\mathbf{A}$ is given by

$$\mathbf{\Omega} = \frac{1}{r}\left\{\mathbf{I} + \nu\mathbf{N}\mathbf{N}' - \mu\mathbf{N}(\mathbf{J}_r \otimes \mathbf{I}_s)\mathbf{N}'\right\} \tag{4.6}$$

4.4 α-Designs

In practice, resolvable designs are required for a wide range of block sizes (k), blocks per replicate (s) and replicate numbers (r). For instance, in agricultural and forestry field trials, efficient designs with low numbers of replicates are usually required. When $k = s$ or $k = s - 1$ square and rectangular lattice designs can be used. The cyclic designs in Chapter 3 also contribute a number of resolvable designs as shown by David (1967). For example, for $v = 12$ and $k = 4$ the cyclic sets with initial blocks (0 1 2 3), (0 1 3 6), (0 1 6 7) and (0 3 6 9) are resolvable. The first two sets have $r = k = 4$, the third $r = 2$ and the last $r = 1$. There remains, however, a large number of combinations of v and r for which no resolvable design exists or where the average efficiency factor of the cyclic design is low. A more general series of resolvable designs for $v = ks$ treatments, called α-designs, has been given by Patterson and Williams (1976a). For these designs there is no limitation on block size other than the unavoidable constraint that v/k must be an integer.

4.4.1 Construction

The construction of an α-design starts with a $k \times r$ generating array α whose elements are in the set of residues modulo s. Each column of α is used to generate $s - 1$ further columns by cyclic substitution. The resulting $k \times rs$ array is denoted by α^*. Finally, s is added to each element in the second row of α^*, $2s$ is added to each element in the third row, and so on. The columns of the resulting array are now the blocks of the required design and each set of columns generated from the same column of α constitutes a replicate.

Each column of the generating array α can also be regarded as an initial block of a GCIB_k design, defined in Section 3.6, if element j in the ith row of the array is replaced by $kj + (i - 1)$.

Example 4.2
Consider the construction of an α-design for three replicates of 12 treatments, each replicate consisting of three blocks of four plots, i.e. $v = 12$, $k = 4$, $r = s = 3$, $b = 9$. Given the generating array α below, the intermediate array α^* and, hence, the design are easily constructed.

Generating array α			Intermediate array α^*								
0	0	0	0	1	2	0	1	2	0	1	2
0	1	2	0	1	2	1	2	0	2	0	1
0	2	1	0	1	2	2	0	1	1	2	0
0	0	1	0	1	2	0	1	2	1	2	0

		Design								
Replicate		1			2			3		
Block	1	2	3	1	2	3	1	2	3
	0	1	2	0	1	2	0	1	2
	3	4	5	4	5	3	5	3	4
	6	7	8	8	6	7	7	8	6
	9	10	11	9	10	11	10	11	9

This design is equivalent to the GCIB_4 design given by the initial blocks (0 1 2 3), (0 5 10 3) and (0 9 6 7).

From the results in Section 3.6, it follows that the dual of an α-design is an α-design. Williams (1975a) shows that the generating array of the dual design is the $r \times k$ array α' with elements $s - \alpha(j, i)$, modulo s, where the elements of the generating array α are $\alpha(i, j)$.

4.4.2 Reduced array

The generating array α in Example 4.2 is of a special kind, called a *reduced* array, with all elements equal to zero in the first row and column. Under a suitable relabelling or permutation of elements all arrays can be represented as reduced arrays, thereby restricting the search for optimal designs to reduced arrays only. For example, consider the following generating array for $s = 3$:

$$
\begin{array}{ccc}
1 & 0 & 1 \\
2 & 2 & 1 \\
0 & 1 & 1 \\
1 & 0 & 2
\end{array}
$$

First add 2 to the elements in columns 1 and 2 to those in column 3, in each case reducing modulo 3 when necessary. This results in a rearrangement of blocks and gives the array

$$
\begin{array}{ccc}
0 & 0 & 0 \\
1 & 2 & 0 \\
2 & 1 & 0 \\
0 & 0 & 1
\end{array}
$$

Now add 2 to the elements in rows 2 and 1 to those in row 3 and again reduce modulo 3 when necessary. This operation simply relabels the treatments and the resulting array is the reduced array used in Example 4.2.

4.4.3 Concurrences

Different α-designs can be obtained from different generating arrays and the choice of an appropriate design can again be based on the efficiency factors of the designs. The problem of choosing optimal designs can be overcome by a combination of theory and computing. The theory can be used to identify classes of designs with high efficiency factors. These designs can then be generated on a computer and compared with upper bounds for the average efficiency factor, given in Section 4.10.

The efficiency factors of a design of a given size depend on the off-diagonal elements of the concurrence matrix $\mathbf{NN}'$. The search for optimal α-designs is simplified by the fact that the number of concurrences of any two treatments, i.e. the number of blocks containing both treatments, can be determined directly from the generating array α. It is not necessary to construct the full design. Let $\alpha(i, j)$ be the (ij)th element of the array α ($i = 1, 2, \ldots, k$; $j = 1, 2, \ldots, r$). Then, the number of concurrences of treatments p and q is the frequency of $(q - p)$ modulo s in the set of r differences $\{\alpha(i_q, j) - \alpha(i_p, j)\}$ modulo s ($j = 1, 2, \ldots, r$), where i_p is one more than the integer part of p/s.

As an illustration, consider the concurrences of treatments 1 and 8 in the design for 12 treatments in blocks of four given in Example 4.2. The values of i_p, i_q and $(q-p)$ modulo 3 are 1, 3 and 1 respectively. The elements of the first row of α have, therefore, to be subtracted from those in the third row. The three differences are 0, 2, 1 and since 1 occurs once then treatments 1 and 8 concur once, namely in the third replicate. It follows immediately that the pairs 0 and 7 and 2 and 6 also concur once. Other pairs of treatments also concur once but some concur twice and others not at all.

A design with concurrences $g_1, g_2, \ldots$ will be referred to as an $\alpha(g_1, g_2, \ldots)$-design; the design in Example 4.2 is therefore an $\alpha(0,1,2)$-design.

4.5 Choice of α-design

Designs with maximum average efficiency factor E among all α-designs are called *α-optimal*. For $r = 2$, as will be shown in Section 4.7, the α-optimal designs can be obtained from the most efficient symmetric cyclic designs for s treatments in blocks of k. For a few combinations of v, k and $r = 3$ or 4 the α-optimal designs are known, for instance when they correspond to square lattice designs. In general, however, a search for the most efficient designs will be necessary.

Efficient α-designs can be obtained using the criteria given in Section 2.5:

(i) For $r \geq k$ the set of α-designs with high efficiency factors are those which minimize the range of off-diagonal elements in the concurrence matrix.

(ii) For $r < k$ the concurrence matrix $\mathbf{N'N}$ of the dual design is smaller than $\mathbf{NN'}$ and so it is better to concentrate on the elements of $\mathbf{N'N}$ and use (2.22).

In both these cases an $\alpha(0,1)$-design would be preferred to an $\alpha(0,1,2)$-design. If an $\alpha(0,1,2)$-design has to be used then one which has as few pairs of treatments as possible concurring twice would be preferred.

From the construction of α-designs it is clear that neither balanced α-designs nor $\alpha(1,2)$-designs can exist since the concurrences of any two treatments i and j such that the integer parts of i/s and j/s are equal is impossible. For instance, in the design for 12 treatments in Example 4.2 it is not possible for treatments 0 and 1 to occur together in any block.

A necessary, though not sufficient, condition for the existence of an $\alpha(0,1)$-design is that $k \leq s$. Williams (1975a) lists an $\alpha(0,1)$-design for 188 of the 198 combinations of v, r and k satisfying $k \leq s, r = 2, 3, 4$ and $v \leq 100$. Of the remaining ten cases, lattice designs exist for nine of them while an $\alpha(0,1,2)$-design has to be used for the case with $s = k = 6$ and $r = 4$.

The tables of designs given by Williams (1975a) are, however, large and fairly inaccessible. As an alternative, Patterson, Williams and Hunter (1978) have given 11 basic arrays which can be used to provide α-designs with k equal to the smaller of s and the integer part of $100/s$, s ranging from 5 to 15 and $r = 2, 3$ or 4. The arrays must not be used to obtain designs with $k < 4$.

A further alternative to these tables of designs is the computer algorithm for choosing suitable generating arrays described by Paterson and Patterson (1983). The choice of array is based on the aim of minimizing the number of circuits of lengths 3 and 4 in an α-design; the use of circuits to obtain efficient block designs has been discussed in Section 2.5. One obvious benefit of the algorithm is that it provides a method of obtaining efficient α-designs for parameter values not included in the tables of Williams.

The algorithm is incorporated into the design generation package ALPHA+ (Williams and Talbot, 1993). This package produces randomized layouts for α-designs for parameters in the ranges

$$2 \leq r \leq 10 \quad 2 \leq k \leq 20 \quad v \leq 500$$

Designs generated by ALPHA+ usually have average efficiency factors within 1% of an upper bound (Section 4.10) and so are suitable for use in practice.

ALPHA+ also has facilities to produce efficient Latinized designs, designs with unequal block sizes and resolvable row–column designs; see Sections 4.6, 4.9 and 6.3.

4.6 Latinized block designs

So far the types of resolvable block designs constructed have had a nested blocking structure, i.e. the incomplete blocks within each replicate are randomized separately for each replicate. Such a randomization process is appropriate for an experimental layout where the replicates are separate entities. Then the blocks of one replicate have no relation to those of another.

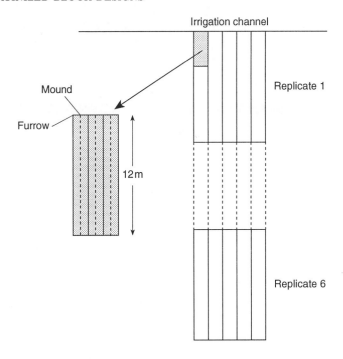

Figure 4.1. *Layout of a cotton variety trial*

In many practical situations, however, the replicates are next to each other or *contiguous*. Consider, for example, the field layout of annual cotton variety trials in Australia. The experimental site is laser planed and furrows one metre apart are cut, running away from an irrigation channel. Each plot consists of three mounds of cotton bushes, of which only the centre mound is used for analysis. Field shape limitations plus possible large furrow-to-furrow variation make it desirable to restrict the number of furrows used in a variety trial; replicates are therefore placed under each other as shown in Fig. 4.1.

If the columns in each replicate correspond to incomplete blocks, then separate block randomizations for each replicate could result in the same variety (treatment) appearing more than once in a (long) block of plots. This would not be desirable if, for example, the irrigation failed on that long block. To cater for this type of situation in the case of $v = s(s - 1)$ treatments, Harshbarger and Davis (1952) introduced *Latinized* rectangular lattice designs.

In these designs there are r replicates and s long blocks with incomplete blocks of $s - 1$ plots in each replicate. For example, consider the following rectangular lattice design for $s = 4$ and $k = 3$:

Replicate				
1	0	1	2	3
	4	5	6	7
	8	9	10	11
2	1	0	3	2
	6	7	4	5
	11	10	9	8
3	2	3	0	1
	7	6	5	4
	9	8	11	10
4	3	2	1	0
	5	4	7	6
	10	11	8	9

With the replicates set out next to each other, the design can also be viewed as a block design with four long blocks of size 12. Further, each treatment occurs once in each long block. Thus an extra design feature has been incorporated into the basic rectangular lattice design to make the layout more appropriate for the situation where replicates are contiguous.

This concept can be extended to cover any resolvable design for $v = ks$ treatments with r replicates. The additional requirement is that the design with s long blocks of size rk should also be highly efficient. Williams (1986a) discussed designs in which the treatments do not occur more than once in each long block, i.e. the long block design is binary. Since there are rk plots in a long block for $v = ks$ treatments, a binary long block design requires $r \leq s$. In practice this would normally be seen as a desirable restriction to maintain a compact structure for the experiment. Contiguous replicate designs can, however, be considered with $r > s$ by using an (M,S)-optimal non-binary block design in which the treatments appear in long blocks either i or $i + 1$ times where i is the integer part of r/s.

Latinized block designs introduce an extra blocking structure into the intra-block analysis of variance table, namely the long blocks. This can have considerable advantage in terms of increasing the precision of estimated treatment means. Further discussion is given in the context of Latinized row–column designs in Section 6.5.

4.6.1 Construction

A convenient method for constructing Latinized block designs with $r \leq s$ is to use α-designs. It will be necessary to place restrictions on the elements $\alpha(i, j)$ of an α-design generating array. Consider the following example:

Example 4.3
For $v = 24$, $r = k = 4$, $s = 6$ the generating array

$$
\begin{array}{cccc}
0 & 1 & 2 & 5 \\
0 & 3 & 5 & 4 \\
0 & 4 & 1 & 3 \\
0 & 5 & 3 & 1 \\
\end{array}
$$

leads to the α-design

Replicate						
1	0	1	2	3	4	5
	6	7	8	9	10	11
	12	13	14	15	16	17
	18	19	20	21	22	23
2	1	2	3	4	5	0
	9	10	11	6	7	8
	16	17	12	13	14	15
	23	18	19	20	21	22
3	2	3	4	5	0	1
	11	6	7	8	9	10
	13	14	15	16	17	12
	21	22	23	18	19	20
4	5	0	1	2	3	4
	10	11	6	7	8	9
	15	16	17	12	13	14
	19	20	21	22	23	18

with the Latinized property that no treatment appears more than once in any long block.

In general if

$$\alpha(i,j) - \alpha(i,j') \neq 0 \quad \text{(modulo } s\text{)} \qquad (4.7)$$

$$(j \neq j'; \ 1 \leq i \leq k; \ 1 \leq j, j' \leq r)$$

the resulting α-design will have the Latinized property. This can be seen by considering, say, the elements $\alpha(1,1)$ and $\alpha(1,2)$ of the generating array. From Section 4.4.1 it follows that $\alpha(1,1)$ generates the treatments $\alpha(1,1) + j$ modulo s in block j of replicate 1 ($j = 0, 1, \ldots, s-1$). The element $\alpha(1,2)$ generates the treatments $\alpha(1,2) + j$ modulo s in block j of replicate 2. Hence long block j will have the same treatment in replicates 1 and 2 if and only if $\alpha(1,1) = \alpha(1,2)$ and so (4.7) follows. It can be seen that the generating array in Example 4.3 satisfies (4.7) since the elements in each row of the array are distinct.

The software package ALPHA+ provides a facility to generate a wide range of efficient Latinized α-designs for use in practice.

4.6.2 Optimality

The Latinized property for a contiguous replicate design is very desirable but it can involve a cost in terms of the optimality of the resolvable incomplete block design. For example, take any two replicates of the square lattice design for nine treatments given in Section 4.1; it is not possible to order the blocks to produce a Latinized block design. Since the square lattice design is optimal (see Section 4.10), it follows that the average efficiency factor for the optimal Latinized block design will be less than that for the square lattice design with the same parameters.

More generally the block concurrence matrix $\mathbf{N'N}$ of a Latinized block design can be partitioned into r^2, $s \times s$ submatrices such that the off-diagonal submatrices have zero elements along the lead diagonal. Hence from (4.1) it is clear that a square lattice design cannot be a Latinized block design. In fact by applying the (M,S)-optimality arguments of Section 2.5 to $\mathbf{N'N}$, it follows that when $k \geq s$ a Latinized block design cannot be optimal.

When $k < s$ and $r \leq s$ it is usually possible to convert an optimal α-design into a Latinized α-design with the same average efficiency factor. While it is clear from (4.7) that a reduced α-array cannot yield a Latinized α-design, column operations similar to those carried out in Section 4.4.2 can usually be carried out on the array to produce a Latinized α-array. For example, by adding 1 to column 2 and 2 to column 3 of the reduced array

$$\begin{array}{ccc} 0 & 0 & 0 \\ 0 & 1 & 2 \\ 0 & 3 & 1 \end{array}$$

for $s = 5$, the array

$$\begin{array}{ccc} 0 & 1 & 2 \\ 0 & 2 & 4 \\ 0 & 4 & 3 \end{array}$$

is obtained which satisfies (4.7) and so will produce a Latinized α-design with the same average efficiency factor as the α-design generated from the reduced array.

4.7 Two-replicate designs

Resolvable block designs with two replicates are important in practice when economy is required in the amount of experimental material (seed, for example) or in the number of plots. Patterson and Williams (1976b) present results which enable the optimal resolvable designs for $r = 2$ to be determined. They show that every binary resolvable incomplete block design for $v = ks$, $b = 2s$ and $r = 2$ is uniquely determined by a symmetric block design with k replications of s treatments. This symmetric design is called the *contraction* of the resolvable design.

An example will illustrate the general approach to establishing this result. Consider the following two-replicate resolvable design for 12 treatments in eight blocks of three, where the two replicates are labelled 0 and 1 and, within each replicate, blocks are labelled 0, 1, 2, 3:

Replicate	0				1			
Block	0	1	2	3	0	1	2	3
	0	1	2	3	0	1	2	3
	4	5	6	7	5	6	7	4
	8	9	10	11	10	11	8	9

The *block concurrence* graph for this design is shown in Fig. 4.2. A block concurrence graph, based on the block concurrence matrix $\mathbf{N'N}$, has the blocks of the design as points and the lines correspond to treatments which pairs of blocks have in common. For instance, there is a line between block 1 in replicate 0 and block 3 in replicate 1 showing that these two blocks have a treatment in common, namely treatment 9.

Blocks in replicate 0

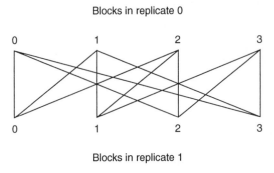

Blocks in replicate 1

Figure 4.2. *The block concurrence graph*

The graph can also be regarded as a *design* graph. Now the four upper points correspond to four treatment labels and the four lower points to four block labels. If a line joins a block point to a treatment point then the block will contain that treatment. Hence, the design graph produces the symmetric block design (contraction) with four treatments in four blocks of three given by

$$
\begin{array}{cccc}
0 & 1 & 2 & 3 \\
1 & 2 & 3 & 0 \\
2 & 3 & 0 & 1
\end{array}
$$

The same graph can, therefore, be regarded as giving rise to both the resolvable design and its contraction, hence establishing the correspondence between them. Patterson and Williams (1976b) proved that the average efficiency factor E of the resolvable design is given by

$$
E = \frac{v - 1}{v - 1 - 2(s - 1) + 4(s - 1)E^{*-1}} \tag{4.8}
$$

where E^* is the average efficiency factor of the contraction.

An immediate consequence of (4.8) is that the resolvable design is A-optimal if the contraction is A-optimal. The above contraction with four treatments, for example, is a balanced incomplete block design (it is also a cyclic design) so that the resolvable design given for 12 treatments in eight blocks of three is A-optimal.

If the contraction is a complete block design ($k = s$) the resulting two-replicate resolvable design is a simple lattice design, which is therefore A-optimal. If the contraction is a balanced incomplete block design the resolvable design is also A-optimal; included here are the simple rectangular lattice designs. If $k > s$ the contraction will be non-binary. Consideration of the counting rules given in Section 2.5 shows that the incidence matrix $\mathbf{N}$ of an efficient non-binary design will be of the form $\mathbf{N} = \mathbf{N}_0 + p\mathbf{J}$, for some integer $p > 1$ and where $\mathbf{N}_0$ is the incidence matrix of a binary design, i.e. within each block some treatments are replicated p times while the others are replicated $p + 1$ times. Again, if $\mathbf{N}_0$ is the incidence matrix of a balanced incomplete block design the resolvable design will be A-optimal.

If the contraction is a symmetric cyclic design then the two-replicate resolvable design will be an α-design. In particular if the cyclic contraction has initial block $(0 \ a_1 \ a_2 \ \dots \ a_{k-1})$ then the two-replicate resolvable design is an α-design obtained from the generating array

$$
\begin{array}{cc}
0 & 0 \\
0 & a_1 \\
0 & a_2 \\
\vdots & \vdots \\
0 & a_{k-1}
\end{array}
$$

Hence, as stated in Section 4.5, the α-optimal designs are obtained from the most efficient symmetric cyclic designs for s treatments in blocks of k. The cyclic designs required for this purpose can be obtained from the tables or computer algorithms referred to in Section 3.4.

Example 4.4

Consider the construction of a two-replicate α-design for 54 treatments in blocks of nine. The most efficient symmetric cyclic set for six treatments in blocks of three has initial block $(0 \ 1 \ 3)$, so that an efficient non-binary cyclic contraction for $s = 6$ and $k = 9$ is obtained from the initial block $(0 \ 1 \ 2 \ 3 \ 4 \ 5 \ 0 \ 1 \ 3)$. Hence, the generating array of the α-design is given by the transpose of

$$
\begin{array}{ccccccccc}
0 & 0 & 0 & 0 & 0 & 0 & 0 & 0 & 0 \\
0 & 0 & 1 & 1 & 2 & 3 & 3 & 4 & 5
\end{array}
$$

The full α-design can now be set out following the method of construction given in Section 4.4.1.

If two-replicate Latinized α-designs are required then it is necessary to choose a cyclic contraction whose initial block does not contain the element 0. When $k < s$ this is easily organized by adding an appropriate number modulo s to each element of an optimal initial block, similar to what was done in Section 4.6.2. If $k \geq s$ then a sub-optimal contraction must be chosen to satisfy (4.7). Further results on the construction of optimal two-replicate resolvable designs can be found in Williams, Patterson and John (1976, 1977).

4.8 Paired comparison designs

The dual of a two-replicate α-design is a resolvable paired comparison $(k = 2)$ design. In fact, Williams (1976) first established the link between the average efficiency factor E of an α-design with $k = 2, v = 2s$ and generating array

$$
\begin{array}{cccc}
0 & 0 & \ldots & 0 \\
0 & a_1 & \ldots & a_{r-1}
\end{array}
$$

and the average efficiency factor E^* of a cyclic design for s treatments with initial block $(0\ a_1\ \ldots\ a_{r-1})$. Using (4.8) and (2.22) it can be shown that E is given by (4.8) with $k = 2$. Hence, α-optimal designs for $k = 2$ are obtained from the most efficient cyclic designs. Many A-optimal resolvable paired comparison designs can be constructed this way as many of the A-optimal block designs are cyclic.

For $6 \leq v \leq 14, r \leq s$ and $k = 2$ Williams (1976) compared the best α-designs with the best resolvable cyclic designs given by David (1967). In 18 of the 20 cases considered the α-designs were either better than or equivalent to the cyclic designs. He also showed that the best α-designs compare favourably with the paired comparison designs listed in the catalogue of cyclic designs by John, Wolock and David (1972), even though these cyclic designs are not necessarily resolvable.

4.9 Unequal block sizes

Resolvable designs with equal sized blocks of k plots with $k < v$ can be used only when v is a multiple of k. In many practical cases a factorization of the number of treatments in the form $v = ks$ may not be possible. But because of the importance of resolvability in practice, it becomes necessary to use unequal block sizes.

Patterson and Williams (1976a) consider the use of two block sizes k_1 and k_2 with the inequality in block size minimized by imposing the condition that $k_2 = k_1 - 1$. By expressing v in the form $v = s_1 k_1 + s_2 k_2$, where s_1, s_2, k_1 and k_2 are positive integers, the designs are derived as follows:

(i) Construct an α-design for $v + s_2$ treatments with $s = s_1 + s_2$ blocks of k_1 plots in each replicate.

(ii) Delete a set of s_2 treatments, no two of which concur. The varieties labelled $v, \ldots, v + s_2 - 1$ provide such a set.

For example, deletion of the treatments labelled 10 and 11 from the design in Example 4.2 gives a design for ten treatments with one block of four plots and two blocks of three plots in each of three replicates.

The α-designs chosen by the above method are in fact optimal or near optimal for $v + s_2$ treatments in blocks of size k_1, but Patterson and Williams (1976a) argue that derived designs with unequal block sizes are also likely to be very efficient in the class of resolvable incomplete block designs with blocks of k_1 and k_2 plots.

Designs with unequal block sizes are used extensively in practice (Patterson and Silvey, 1980). One difficulty, however, is the specification of the variance matrix for analysis with recovery of inter-block information; this is discussed in Section 7.7.

4.10 Upper bounds for the average efficiency factor

In order to compare alternative resolvable designs and to provide measures against which to judge the best available designs it is valuable to have upper bounds for the average efficiency factor E. When using computer search algorithms, such as those in ALPHA+, the algorithm can be stopped when the best design found is sufficiently close to the bound. The upper bounds in Section 2.8 can be used for resolvable block designs. For $v > b$, however, it is usually possible to develop better upper bounds for E by making use of structure in the block concurrence matrix $\mathbf{N'N}$. Hence in this section some upper bounds are derived by first finding bounds for the dual of a resolvable design and substituting into (2.22). The best upper bound for E is then the lowest of all the available upper bounds.

The block concurrence matrix $\mathbf{N'N}$ for a resolvable design with $v = ks$ treatments in rs blocks of k plots per block can be partitioned into $s \times s$ submatrices $\mathbf{B}_{ij}$, say $(i, j = 1, 2, \ldots, r)$. The (ij)th element of $\mathbf{N'N}$ gives the number of treatments common to both the ith and jth blocks of the design. Two different blocks in the same replicate will have no treatments in common so that $\mathbf{B}_{ii} = k\mathbf{I}_s$ $(i = 1, 2, \ldots, r)$. Consider the s blocks in ith replicate and one block from the jth replicate $(i \neq j)$. Each of the k treatments from this jth replicate block must occur once, and only once, in one of the blocks in the ith replicate. Hence, the sum of each row of $\mathbf{B}_{ij}$ is equal to k, i.e. $\mathbf{B}_{ij}\mathbf{1} = k\mathbf{1}$ $(i, j = 1, 2, \ldots, r; \ i \neq j)$. Note that for square lattice designs $\mathbf{B}_{ij} = \mathbf{J}_s$ from (4.1) and for rectangular lattice designs $\mathbf{B}_{ij} = \mathbf{J}_s - \mathbf{I}_s$ from (4.4), for all $i \neq j$.

It follows that $(r - 1)$ eigenvalues of $\mathbf{N}'\mathbf{N}$ are equal to zero. These correspond to the $(r - 1)$ linearly independent eigenvectors representing replicate contrasts. One such eigenvector, for instance, is given by $(\mathbf{1}'_s, -\mathbf{1}'_s, \mathbf{0}'_s \ldots \mathbf{0}'_s)'$. Hence, the canonical efficiency factors of the dual of a resolvable design are

$$e_1 = e_2 = \ldots = e_{r-1} = 1, \quad e_r, \ldots, e_{b-1}$$

Their sum, given by the trace of $\mathbf{I} - (1/rk)\mathbf{N}'\mathbf{N}$, is $s(r - 1)$ so that

$$\sum_{i=r}^{b-1} e_i = (s - 1)(r - 1) \tag{4.9}$$

Using (2.22) an upper bound for E can be obtained as

$$U = \frac{v - 1}{v - 1 - r(s - 1) + r(s - 1)U_r^{-1}} \tag{4.10}$$

where U_r is an upper bound for the harmonic mean of $e_r, \ldots, e_{b-1}$. Analogous to $\bar{e}$ in (2.26) an upper bound for a resolvable design with $v \geq b$ can be obtained by replacing U_r by the arithmetic mean of the $r(s-1)$ eigenvalues $e_r, \ldots, e_{b-1}$, namely $\bar{e}_r = (r-1)/r$. After simplification this gives the upper bound

$$U_{0r} = \frac{(v - 1)(r - 1)}{(v - 1)(r - 1) + r(s - 1)} \tag{4.11}$$

This result was given by Patterson and Williams (1976b). Note that this bound is attained by square lattice designs; see (4.2).

Tighter values for U can be obtained by applying the methods of Section 2.8 to U_r. For example, Jarrett (1989) defines S_{2r} and S_{3r} as the corrected second and third moments of the $r(s - 1)$ eigenvalues $e_r, \ldots, e_{b-1}$. A lower bound for S_{2r} occurs when the elements of $\mathbf{B}_{ij}$ $(i \neq j)$ differ by at most one. This gives

$$S_{2r} \geq S_{2rL} = r(r - 1)s^2 \beta(1 - \beta)/(rk)^2$$

where β is the fractional part of k/s. An upper bound for E, given by Williams and Patterson (1977), can then be written by replacing

U_r with U_{1r} in (4.10) where

$$U_{1r} = \bar{e}_r - \frac{\bar{e}_r S_{2rL}}{r(s-1)\bar{e}_r + S_{2rL}}$$

Another upper bound is obtained by replacing $\bar{e}$, S_2 and $v-1$ by $\bar{e}_r$, S_{2rL} and $r(s-1)$ respectively in (2.29) to give U_{2r}, say, and then replacing U_r with U_{2r} in (4.10). This bound corresponds to that described in Proposition 3.1 of Tjur (1990). It has been shown by Williams and John (1993) that using U_{2r} gives a tighter bound than with U_{1r}.

Analogous to (2.31), a third moment upper bound for the harmonic mean of $e_r, \ldots, e_{b-1}$ can be written as

$$U_{3r} = \bar{e}_r - \frac{S_{2r}^2}{r(s-1)(S_{3r} + \bar{e}_r S_{2r})} \qquad (4.12)$$

Under the assumption that the elements of $\mathbf{B}_{ij}$ ($i \neq j$) differ by at most one, Williams and Patterson (1977) and Jarrett (1989) give the following bound for S_{3r}:

$$S_{3r} \geq S_{3rW} = r(r-1)(r-2)s^2\beta x/(rk)^3 \qquad (4.13)$$

where

$$x = \begin{cases} s\beta^2 & \text{if } \beta \leq 1/2 \text{ and } k \geq s \\ s\beta^2 - 1 & \text{if } \beta \leq 1/2 \text{ and } k < s \\ s(1-\beta)^2 & \text{if } \beta > 1/2 \end{cases}$$

Substituting S_{2rL} and S_{3rW} into (4.12) and replacing U_r with U_{3r} in (4.10) gives a bound which is conjectured to be an upper bound for E. Jarrett (1989) showed that this bound is tighter than a third moment bound derived by Williams and Patterson (1977).

Example 4.5
Suppose an upper bound on E is required for resolvable block designs with $v = 40$, $k = 5$, $s = 8$ and $r = 3$. From (4.11), $U_{0r} = 0.7879$ but using U_{2r} and U_{3r} leads to upper bounds 0.7725 and 0.7711 respectively. The best upper bound from Section 2.8 is 0.7725, so in this case the third moment bound applied to the dual design gives the tightest bound.

For $r = 2$ the odd moments S_{3r}, S_{5r} ... are zero and so the bound obtained from (4.12) can usually be improved on. The most convenient approach is to use the results of Section 4.7 and define upper bounds on the contraction of a two-replicate resolvable block design. Following (4.8) an upper bound for the average efficiency factor of a two-replicate resolvable block design is given by

$$U = \frac{v-1}{v - 1 - 2(s-1) + 4(s-1)U^{*-1}} \tag{4.14}$$

where U^* is an upper bound for the average efficiency factor of the contraction, obtained from Section 2.8.

Example 4.6
For two-replicate resolvable block designs with $v = 54$, $k = 9$ and $s = 6$, $\bar{e}^* = 0.9778$ and so $U_2^* = 0.9776$ giving an upper bound for E of 0.8352. In this case substitution of $\bar{e}^*$ for U^* in (4.14) also gives the same bound to four decimal places.

Upper bounds for the average efficiency factor of Latinized block designs have been developed by Williams and John (1993). As mentioned in Section 4.6.2, for $k \geq s$ the Latinized property represents a reduction in E and so the Latinized bounds allow a more accurate assessment of the possible improvement in a design search algorithm. For a Latinized block design the B_{ij} $(i \neq j)$ can be written so that the diagonal elements are zero. Hence a lower bound for S_{2r} occurs when the off-diagonal elements of B_{ij} $(i \neq j)$ differ by at most one. This gives

$$S_{2r} \geq S_{2lL} = \frac{r(r-1)\{k^2 + s(s-1)^2\gamma(1-\gamma)\}}{(s-1)(rk)^2}$$

where γ is the fractional part of $k/(s-1)$.
    For B_{ij} of the above form

$$S_{3r} \geq S_{3lW} = r(r-1)(r-2)y^*/(rk)^3$$

where

$$y^* = \begin{cases} y & \text{if } \gamma \leq \gamma_0 \text{ and } k \geq s - 1 \\ y - s(s-1)\gamma & \text{if } \gamma \leq \gamma_0 \text{ and } k < s - 1 \\ y - s(s-1)\gamma(2s\gamma - 2\gamma - s) & \text{if } \gamma > \gamma_0 \end{cases}$$

and where

$$y = s(s-1)(s+1)\gamma^3 - 3s(s-1+k)\gamma^2 + 3sk\gamma + k^3/(s-1)^2$$

and $\gamma_0 = s/\{2(s-1)\}$.
    The quantities S_{2lL} and S_{3lW} can be used in a similar manner to S_{2rL} and S_{3rW} previously in this section, to give a second moment and a conjectured third moment bound for Latinized block designs.

Example 4.7

The best upper bound on E for resolvable Latinized block designs with $v = 24$, $k = 6$, $s = 4$ and $r = 3$ is 0.8163, which is the third moment bound obtained using S_{2lL} and S_{3lW}. This is tighter than those for non-Latinized resolvable and non-resolvable designs, which are 0.8298 and 0.8451 respectively.

CHAPTER 5

Row–column designs

5.1 Introduction

In row–column designs the experimental units are grouped in two directions, i.e. two blocking factors are used with one factor representing the rows of the design and the other factor representing columns. For instance, in an agricultural experiment the rows and columns of a design might be used to control field gradient and soil type respectively. In a preference testing experiment, the columns might represent different subjects and the rows the order in which different products are presented to the subjects. In an insulin-response experiment, the rows, columns and treatments may represent respectively individual animals, date of injection and levels of insulin. In all these experiments it is hoped that there will be a gain in the accuracy of estimating treatment comparisons resulting from eliminating the effects of the row and column factors.

The designs considered in this chapter have v treatments set out in an array of k rows and s columns such that the ith treatment is replicated r_i times. For example, a (Latin square) design for $v = 4$ treatments, labelled 0, 1, 2 and 3, in $k = 4$ rows and $s = 4$ columns is given by

$$\begin{matrix} 0 & 1 & 2 & 3 \\ 1 & 3 & 0 & 2 \\ 2 & 0 & 3 & 1 \\ 3 & 2 & 1 & 0 \end{matrix}$$

where each treatment is replicated four times.

The degrees of freedom in a row–column design can be partitioned into three blocking strata as shown in Table 5.1. Information on treatment comparisons may be available in all strata, though comparisons made in the lowest stratum, namely the within rows and columns stratum, will usually have the greatest precision. The aim in choosing an appropriate row–column design

Table 5.1. *Partition of degrees of freedom in a row–column design*

Stratum	d.f.
Between rows	$k - 1$
Between columns	$s - 1$
Within rows and columns	$(k - 1)(s - 1)$
Total	$ks - 1$

will, therefore, be to maximize the amount of information available on treatment comparisons in this stratum.

More generally, an experiment may be made up of groups of row–column designs, each with k rows and s columns. These groups may, for instance, represent a further blocking factor or provide repeats of the basic row–column design. For example, the fertilizer experiment referred to above may be repeated in various locations. Of particular importance are resolvable row–column designs, where each treatment occurs once in each row–column design and the groups correspond to replicates. These designs will be considered in detail in Chapter 6.

5.2 Model and information matrix

Let y_{ijm} be the response obtained when the ith treatment is applied in the jth row and mth column. Then the model corresponding to the partition given in Table 5.1 is

$$y_{ijm} = \mu + \tau_i + \rho_j + \gamma_m + \varepsilon_{ijm} \qquad (5.1)$$

$$(i = 1, 2, \ldots, v; j = 1, 2, \ldots, k; m = 1, 2, \ldots, s)$$

where the parameters τ_i, ρ_j and γ_m measure the effect of the ith treatment, jth row and mth column respectively. The error terms ε_{ijm} are assumed to be uncorrelated random variables each with mean zero and variance σ^2.

Let D_k and D_s denote the block designs given by the rows and columns respectively of the row–column design. For instance, in the Latin square design for four treatments given in the previous section D_k is a design with four blocks and with each treatment occurring once in each block, i.e. D_k is a complete block design; as is D_s. The block designs D_k and D_s are called the *row component* and *column component* designs respectively of the row–column design. Let $\mathbf{N}_k$ and $\mathbf{N}_s$ be the incidence matrices of D_k and D_s respectively.

The reduced normal equations for the treatment parameters, after eliminating the row and column parameters, can be obtained in a way analogous to the method used in Section 1.4 for block designs. For row–column designs these equations are

$$\mathbf{A}\hat{\tau} = \mathbf{q} \tag{5.2}$$

where the information matrix $\mathbf{A}$ is given by

$$\mathbf{A} = \mathbf{r}^\delta - (1/s)\mathbf{N}_k\mathbf{N}_k' - (1/k)\mathbf{N}_s\mathbf{N}_s' + (1/ks)\mathbf{rr}' \tag{5.3}$$

and the vector of adjusted treatment totals $\mathbf{q}$ by

$$\mathbf{q} = \mathbf{T} - (1/s)\mathbf{N}_k\mathbf{R} - (1/k)\mathbf{N}_s\mathbf{C} + \bar{y}\mathbf{r} \tag{5.4}$$

and where $\mathbf{r}$ is the replication vector and $\mathbf{T}$, $\mathbf{R}$ and $\mathbf{C}$ are respectively vectors of treatment, row and column totals.

The results given in Section 1.4–1.6 for block designs can be applied in the same way to these row–column designs. For instance, a solution to (5.2) is given by $\hat{\tau} = \boldsymbol{\Omega}\mathbf{q}$ where $\boldsymbol{\Omega}$ is a generalized inverse of $\mathbf{A}$; the estimable function $\mathbf{c}'\tau$ satisfies $\mathbf{c}' = \mathbf{c}'\boldsymbol{\Omega}\mathbf{A}$ and has unbiased estimator $\mathbf{c}'\hat{\tau}$ with variance $\mathbf{c}'\boldsymbol{\Omega}\mathbf{c}\sigma^2$; and the (adjusted) treatment sum of squares is $\hat{\tau}'\mathbf{q}$.

5.3 Canonical efficiency factors

Following Section 2.6, let e_i be an eigenvalue corresponding to an eigenvector $\mathbf{x}_i$ of the information matrix $\mathbf{A}$ given in (5.2) with respect to $\mathbf{r}^\delta$, i.e. $\mathbf{A}\mathbf{x}_i = e_i\mathbf{r}^\delta\mathbf{x}_i$ $(i = 1, 2, \ldots, v)$. The non-zero eigenvalues $e_1, e_2, \ldots, e_t$ are the canonical efficiency factors of the row–column design, where $t = \mathrm{rank}(\mathbf{A})$. Optimality criteria based on these factors can be used, as was done with block designs, to assess different row–column designs.

In general, from (2.19), the canonical efficiency factors are most easily obtained as the (non-zero) eigenvalues of the matrix

$$\mathbf{A}^* = \mathbf{r}^{-\delta/2}\mathbf{A}\mathbf{r}^{-\delta/2} \tag{5.5}$$

However, one case of particular interest arises when the information matrices $\mathbf{A}_k$ and $\mathbf{A}_s$ of the component designs are spanned by the same set of eigenvectors. Designs with this property are called *generally balanced* by Houtman and Speed (1983). Suppose

$$\mathbf{A}_k\mathbf{x}_i = e_{ki}\mathbf{r}^\delta\mathbf{x}_i, \qquad \mathbf{A}_s\mathbf{x}_i = e_{si}\mathbf{r}^\delta\mathbf{x}_i$$

Then, since (5.3) can be written as

$$\mathbf{A} = \mathbf{A}_k + \mathbf{A}_s - \mathbf{r}^\delta + (1/ks)\mathbf{r}\mathbf{r}'$$

the ith canonical efficiency factor e_i of the row–column design is given by

$$e_i = e_{ki} + e_{si} - 1 \qquad (5.6)$$

For the equal replication case, (5.6) will hold if the component designs are complete block, balanced incomplete block (or any variance-balanced design), cyclic or n-cyclic designs. Cyclic and n-cyclic designs satisfy this requirement since circulant and block circulant matrices of a given size have a common set of eigenvectors. Note that if one of the component designs is variance-balanced (or efficiency-balanced in the unequal replication case) then (5.6) will hold no matter what type of design is used for the other component.

5.4 Orthogonality and connectedness

In an orthogonal row–column design estimates of the treatment parameters will be the same as those obtained from the model with no row or column parameters. The requirements for orthogonality are, following the methods used in Section 1.9, that

$$\mathbf{N}_k = (1/k)\mathbf{r}\mathbf{1}', \qquad \mathbf{N}_s = (1/s)\mathbf{r}\mathbf{1}' \qquad (5.7)$$

The Latin square design for four treatments given in Section 5.1 is an orthogonal row–column design since $\mathbf{N}_k = \mathbf{N}_s = \mathbf{1}\mathbf{1}'$ satisfies (5.7). The component designs D_k and D_s are both (orthogonal) complete block designs as each treatment occurs once in each row and once in each column.

Using (5.7), the information matrix and vector of adjusted treatment totals for an orthogonal design are, from (5.3) and (5.4),

$$\mathbf{A} = \mathbf{r}^\delta - (1/ks)\mathbf{r}\mathbf{r}' \qquad (5.8)$$

and

$$\mathbf{q} = \mathbf{T} - \bar{y}\mathbf{r} \qquad (5.9)$$

These formulae are the same as for an orthogonal block design given in (1.47) and (1.48) so that the results of Section 1.9 apply here also. In particular, since $\boldsymbol{\Omega} = \mathbf{r}^{-\delta}$ is a generalized inverse of $\mathbf{A}$, estimates of the treatment parameters are given by the unadjusted treatment means.

Another useful concept is that of *adjusted orthogonality*, in which rows are orthogonal to columns after adjusting for treatments. It is defined as follows. Consider the model obtained from (5.1) by deleting the column parameters γ_m, namely

$$y_{ijm} = \mu + \tau_i + \rho_j + \varepsilon_{ijm} \qquad (5.10)$$

Now if the estimates of the row parameters ρ_j are the same for both models (5.1) and (5.10) then rows are said to be adjusted orthogonal to columns. Similarly, adjusted orthogonality means that estimates of column parameters will be the same regardless of whether the row parameters are included in the model.

For designs with equal replication, after eliminating the treatment parameters from the normal equations for model (5.1), it can be shown that the necessary and sufficient condition for adjusted orthogonality between rows and columns is that

$$\mathbf{N}_k'\mathbf{N}_s = r\mathbf{J} \qquad (5.11)$$

This means that there are r treatments common to the jth row and mth column of the design, for all j and m. Eccleston and Kiefer (1981) show that the information matrices $\mathbf{A}$, $\mathbf{A}_k$ and $\mathbf{A}_s$ for row–column designs with adjusted orthogonality have a common set of eigenvectors so that the canonical efficiency factors satisfy (5.6). In addition, it can be shown that $r\mathbf{A} = \mathbf{A}_k\mathbf{A}_s$, so that

$$e_i = e_{ki}e_{si} \qquad (5.12)$$

The properties of adjusted orthogonal row–column designs can, therefore, be determined directly from those of its row component and column component designs. The average efficiency factor E of the row–column design can be expressed as a function of the corresponding efficiency factors E_k and E_s of the component designs. Using (5.6) and (5.12),

$$\frac{1}{e_{ki}} + \frac{1}{e_{si}} = \frac{e_{ki} + e_{si}}{e_i}$$
$$= \frac{1 + e_i}{e_i}$$

i.e.

$$e_i^{-1} = e_{ki}^{-1} + e_{si}^{-1} - 1$$

which leads to

$$E = (E_k^{-1} + E_s^{-1} - 1)^{-1} \qquad (5.13)$$

It follows, therefore, from (5.12) and (5.13) that if the two component designs are E-, D- or A-optimal then, among the class of adjusted orthogonal designs, the row–column design will also be E-, D- or A-optimal.

A row–column design is said to be *connected* if the rank of the information matrix $\mathbf{A}$, given in (5.3), is $v - 1$. All contrasts in the treatment parameters will then be estimable. A design will be connected if and only if it has $v - 1$ canonical efficiency factors; calculation of these factors provides a simple way of determining whether a given design is connected or not.

Pearce (1975) gave the following example of a disconnected row–column design for $v = k = s = 4$:

$$
\begin{array}{cccc}
0 & 0 & 2 & 3 \\
0 & 0 & 3 & 2 \\
2 & 3 & 1 & 1 \\
3 & 2 & 1 & 1
\end{array}
$$

The comparison of treatments 0 and 1 is not estimable; this can easily be verified by showing that the estimability condition of Section 1.5 does not hold for this contrast. Note that, for this design, both the row and column components D_k and D_s are connected. This example shows, therefore, that the connectedness of a row–column design cannot necessarily be established from the connectedness of its component designs.

Row–column designs whose canonical efficiency factors satisfy (5.6) will be disconnected if $e_{ki} + e_{si} = 1$ for some i $(1 \leq i \leq v-1)$, even if both components are connected. This is a generalization of a result given by Russell (1976) who showed that if one component was efficiency-balanced with average efficiency factor E_1, say, then the row–column design will be connected if and only if $1 - E_1$ is not a canonical efficiency factor of the other component; see also Russell (1980). Other results on the connectedness of row–column designs can be found in Butz (1982).

For designs satisfying the adjusted orthogonality condition of (5.11) Raghavarao and Federer (1975) showed that the row–column design is connected if and only if the row component and column component designs D_k and D_s are connected. It follows from (5.12) that e_i is not zero if and only if both e_{ki} and e_{si} are not zero $(1 \leq i \leq v - 1)$.

5.5 Upper bounds for the average efficiency factor

The results in this and subsequent sections will be restricted to row–column designs in which every treatment is replicated r times, so that $vr = ks$.

An upper bound for the average efficiency factor E of a row–

column design is given by the arithmetic mean $\bar{e}$ of the canonical efficiency factors. When both component designs are binary, i.e. no treatment occurs more than once in any row or column, this bound is

$$\bar{e} = \frac{ks - k - s + r}{r(v - 1)} \tag{5.14}$$

When at least one component is non-binary then following from (2.34) and (5.3) an improved bound is given by

$$U_1 = 1 - \frac{k'(v - k')}{k^2(v - 1)} - \frac{s'(v - s')}{s^2(v - 1)} \tag{5.15}$$

where $k' = k$ modulo v and $s' = s$ modulo v; a result given by Park and Dean (1990).

For adjusted orthogonal designs it is clear from (5.13) that an upper bound is given by

$$U_2 = (U_k^{-1} + U_s^{-1} - 1)^{-1} \tag{5.16}$$

where U_k and U_s are upper bounds for the average efficiency factors of the row and column component designs respectively. Values for U_k and U_s can be obtained using the results of Section 2.8. Eccleston and McGilchrist (1985) have shown that U_2, given in (5.16), is in fact an upper bound for any row–column design, irrespective of whether it satisfies the adjusted orthogonality condition (5.11) or not.

For a particular set of parameters, the best upper bound is given by the smallest of $\bar{e}$, U_1 and U_2.

Example 5.1
Suppose an upper bound on E is required for row–column designs with $v = 8$, $k = 4$, $s = 6$ and $r = 3$. The arithmetic mean bound gives $\bar{e} = U_1 = 0.8095$. Using the results in Section 2.8, the best bounds for the component designs are $U_k = 0.9492$ and $U_s = 0.8400$. Substituting these values into (5.16) gives $U_2 = 0.8038$.

Example 5.2
For a row–column design with $v = 9$, $k = 6$, $s = 12$ and $r = 8$, i.e. when one component (the row component) is necessarily non-binary, $U_1 = 0.9219$ and $U_2 = 0.9235$, using $U_k = 0.9841$ and $U_s = 0.9375$.

Example 5.3
For $v = 12$, $k = 4$, $s = 15$ and $r = 5$, $U_1 = 0.8073$ and $U_2 = 0.8028$,

using $U_k = 0.9888$ and $U_s = 0.8102$.

5.6 Latin square designs

If the two component designs D_k and D_s are both complete block designs the resulting row–column design is called a Latin square design. Such a design necessarily has the number of treatments equal to the number of rows and columns, i.e. $v = k = s$. An example of a Latin square design with $v = 4$ has been given in Section 5.1. Latin square designs are, therefore, row–column designs having each treatment occurring once in each row and once in each column, and are easily constructed for any value of v.

It follows, from the results in Section 5.4, that Latin square designs are orthogonal row–column designs, so that estimates of treatment parameters are based on treatment means, no adjustment for rows or columns being necessary. Hence, the contrast $\mathbf{c}'\boldsymbol{\tau} = \sum c_i \tau_i$ in the treatment parameters is estimated by $\sum c_i \bar{y}_i$ with variance $(\sigma^2/r)\sum c_i^2$, where $\bar{y}_i$ is the ith treatment mean.

All the information on treatment contrasts is available from comparisons within rows and columns, since the designs are efficiency-balanced with all canonical efficiency factors equal to unity. The analysis of variance for a Latin square is given in Table 5.2 where $\mathbf{T}$, $\mathbf{R}$ and $\mathbf{C}$ are vectors of treatment, row and column totals respectively and G is the overall total.

Orthogonal row–column design can also be obtained by joining together a number of Latin square designs; the number of rows and columns now being a multiple of the number of treatments. For small values of v, it becomes necessary to use more than one Latin square design in order to have sufficient degrees of freedom

Table 5.2. *Analysis of variance for a Latin square*

	d.f.	s.s.
Between rows	$v-1$	$(1/v)\mathbf{R}'\mathbf{R} - G^2/v^2$
Between columns	$v-1$	$(1/v)\mathbf{C}'\mathbf{C} - G^2/v^2$
Between treatments	$v-1$	$(1/v)\mathbf{T}'\mathbf{T} - G^2/v^2$
Residual	$(v-1)(v-2)$	by subtraction
Total	$v^2 - 1$	$\mathbf{y}'\mathbf{y} - G^2/v^2$

in the analysis of variance to adequately estimate the experimental
error σ^2.

Example 5.4
An orthogonal row–column design with $v = 5$, $k = 5$ rows and
$s = 10$ columns is given by

$$
\begin{array}{cccccccccc}
0 & 1 & 2 & 3 & 4 & 0 & 1 & 2 & 3 & 4 \\
1 & 3 & 4 & 2 & 0 & 1 & 4 & 3 & 0 & 2 \\
2 & 0 & 1 & 4 & 3 & 2 & 0 & 1 & 4 & 3 \\
3 & 4 & 0 & 1 & 2 & 3 & 2 & 4 & 1 & 0 \\
4 & 2 & 3 & 0 & 1 & 4 & 3 & 0 & 2 & 1
\end{array}
$$

obtained by joining together two 5 × 5 Latin squares. Each
treatment now occurs twice in each row and once in each column.

5.7 Row-orthogonal designs

When it is impractical to have a complete replicate of
the treatments in each column (or row), incomplete block
arrangements will have to be used. This section will be concerned
with row–column designs in which each treatment occurs equally
often in each row of the design but where some treatments occur
less frequently than others in the columns. Thus, the number of
columns s must be equal to the number of treatments v. The
column component D_s will usually be a binary incomplete block
design. Since the canonical efficiency factors of D_k are all equal to
unity, it follows from (5.6) that the canonical efficiency factors of
these row–column designs are equal to those of D_s, i.e.

$$e_i = e_{si} \quad (i = 1, 2, \ldots, v - 1) \tag{5.17}$$

Hence, choosing an appropriate (e.g. A-optimal) row–column
design is equivalent to choosing an appropriate column component
design D_s.

From (5.3) and (5.4), the information matrix **A** and the vector
of adjusted treatment totals **q** of these row–column designs are
also equal to those of D_s. Estimates of treatment parameters are,
therefore, the same as those obtained from a model in which the
row parameters have been deleted. These row–column designs will
be said to be *row-orthogonal* designs, since rows are orthogonal
to treatments. Row effects are only involved in estimating the
experimental error in the analysis of variance; with the row sum
of squares being of the same form as that given in Table 5.2,

with k rather than v rows. Note that rows and columns are interchangeable, so that column-orthogonal designs are obtained by simply interchanging rows and columns in row-orthogonal designs.

The symmetric (i.e. $r = k$) balanced incomplete block designs of Chapter 1 and cyclic designs of Chapter 3 can be used as the column component design D_s to provide classes of row-orthogonal row–column designs. It is necessary, of course, to be able to re-arrange the treatments within the blocks so that the row component D_k is a complete block design.

A row-orthogonal design in which D_s is a balanced incomplete block design is called a *Youden square*. It is always possible to set out any symmetric balanced incomplete block design as a Youden square.

Example 5.5
Consider the balanced incomplete block design for $v = 5$ treatments in five blocks of four given in Section 1.2. The treatments can be re-arranged within blocks to give the Youden square design

$$
\begin{array}{ccccc}
0 & 4 & 3 & 2 & 1 \\
1 & 0 & 4 & 3 & 2 \\
2 & 1 & 0 & 4 & 3 \\
3 & 2 & 1 & 0 & 4
\end{array}
$$

Note that each treatment occurs once in each row.

Example 5.6
As a further example, a Youden square design for $v = 13$ treatments in $k = 4$ rows and $s = 13$ columns, obtained by re-arranging the balanced incomplete block design of Example 1.2, is

$$
\begin{array}{ccccccccccccc}
0 & 9 & 8 & 6 & 7 & 2 & 4 & 11 & 3 & 5 & 1 & 12 & 10 \\
1 & 3 & 9 & 0 & 10 & 5 & 8 & 6 & 2 & 7 & 12 & 4 & 11 \\
2 & 4 & 7 & 3 & 1 & 10 & 0 & 5 & 11 & 12 & 8 & 6 & 9 \\
9 & 5 & 6 & 10 & 4 & 8 & 11 & 1 & 7 & 0 & 3 & 2 & 12
\end{array}
$$

This design might be used, for example, in the preference tasting experiment of Section 5.1, where it is felt it was unreasonable for subjects to be expected to discriminate between more than four products (treatments).

Youden square designs are, from (1.39) and (5.17), variance-balanced and optimal under all the optimality criteria given in

Section 2.4. Of course, these designs will only exist for certain parameter combinations so that it is necessary to consider designs based on other incomplete block arrangements.

Designs based on cyclical methods of construction provide a flexible class of row-orthogonal designs. A cyclic set with $r = k$ will necessarily have each treatment occurring once in each row, so that all cyclic designs with the number of blocks (i.e. columns) a multiple of the number of treatments are automatically row-orthogonal. The design for $v = 7$ treatments in seven blocks of three given in Section 3.1 illustrates this, although in this case since the design is also a balanced incomplete block design the resulting row–column design is a Youden square. The tables of cyclic designs listed in John, Wolock and David (1972) and Lamacraft and Hall (1982) can be used to provide efficient row-orthogonal designs. Alternatively, such designs can be obtained through the use of the computer algorithms referred to in Section 3.4. The n-cyclic designs of Section 3.5 also have the property that if the number of blocks is a multiple of the number of treatments then the designs are automatically row-orthogonal.

Consider now the generalized cyclic design of Section 3.6 where the $v = mn$ treatments are divided into m groups of n elements using the residue classes, modulo m. The designs are obtained by successive addition of m, modulo v, to be elements of one or more initial blocks. To obtain a symmetric generalized cyclic design, therefore, m initial blocks are generally needed and the design will be row-orthogonal if the elements in the jth position (i.e. jth row) of each of these m blocks belong to different residue classes $(j = 1, 2, \ldots, s)$.

Example 5.7
For $v = 8$ treatments in $m = 2$ groups of $n = 4$ the residue classes are

$$S_0 = \{0 \quad 2 \quad 4 \quad 6\}$$

and

$$S_1 = \{1 \quad 3 \quad 5 \quad 7\}$$

Let one initial block be (0 1 2 4) so that the other initial block must have one element from S_0 and three from S_1. One such block is (0 1 3 7) and, suitably re-arranged, leads to the following row-orthogonal generalized cyclic design with $k = 4$ and $s = 8$:

$$
\begin{array}{cccccccc}
0 & 2 & 4 & 6 & 7 & 1 & 3 & 5 \\
1 & 3 & 5 & 7 & 0 & 2 & 4 & 6 \\
2 & 4 & 6 & 0 & 1 & 3 & 5 & 7 \\
4 & 6 & 0 & 2 & 3 & 5 & 7 & 1
\end{array}
$$

It can be seen, therefore, that the cyclic designs of Chapter 3 provide a means of obtaining a wide choice of row–column designs in which the rows are orthogonal to treatments.

Note that all row-orthogonal designs satisfy the adjusted orthogonality condition of (5.11). For binary designs the condition $\mathbf{N}'_k \mathbf{N}_s = r\mathbf{J}$ means that each column of the design has r treatments in common with each row, which is a condition clearly satisfied by row-orthogonal designs since $r = k$. Thus, (5.17) could also be obtained from (5.12) as $e_{ki} = 1$ for all i.

5.8 Non-orthogonal row–column designs

Suppose now that neither k nor s is a multiple of v. Then it will not be possible to construct row–column designs in which the treatments are orthogonal to either of the component designs, since neither D_k nor D_s can be a complete block design. The treatments in such row–column designs will necessarily be non-orthogonal to both rows and columns. As an example, consider the following non-orthogonal row–column design given by Pearce (1963) for $v = 4$, $k = s = 6$ and $r = 9$:

$$
\begin{array}{cccccc}
0 & 3 & 2 & 1 & 0 & 3 \\
1 & 1 & 3 & 0 & 2 & 2 \\
3 & 2 & 1 & 2 & 3 & 0 \\
2 & 0 & 3 & 0 & 1 & 1 \\
2 & 2 & 1 & 3 & 0 & 0 \\
1 & 3 & 0 & 3 & 1 & 2
\end{array}
$$

It can be shown that this design is variance-balanced. Further both D_k and D_s are also (non-binary) variance-balanced designs, although this requirement is not necessary for the row–column design to be variance-balanced.

Many different methods of constructing non-orthogonal row–column designs have been given. Shrikhande (1951) gave a series of efficient designs for $v = t^2$, $k = t(t + 1)$, $s = t$ and $r = t + 1$, as well as four efficient designs with $(v, k, s) = (6,10,3)$, $(8,14,4)$, $(10,18,9)$, $(15,35,3)$.

Agrawal (1966) obtained a class of adjusted orthogonal row–column designs for $v = k + s - 1$ treatments whose components are the duals of symmetric balanced incomplete block designs; this class has also been considered by Anderson and Eccleston (1985). It follows from (2.22) and (5.13) that these row–column designs are A-optimal. The designs tabulated by Agrawal (1966) are for $(v, k, s) = (10,5,6)$, $(12,4,9)$, $(14,7,8)$, $(15,6,10)$, $(18,9,10)$, $(20,5,16)$, $(22,11,12)$, $(29,6,25)$, while other designs given by Anderson and Eccleston (1985) have $(v, k, s) = (24,9,16)$, $(26,13,14)$, $(30,10,21)$, $(30,15,16)$. It can be shown that these designs have two distinct canonical efficiency factors given by $(s - 1)/s$ with multiplicity $k - 1$ and $(k - 1)/k$ with multiplicity $s - 1$.

John and Eccleston (1986) gave a class of adjusted orthogonal row–column designs based on the α-designs given in Section 4.4. Designs given earlier by Raghavarao and Shah (1980) are a special case of these designs. Lewis and Dean (1991) have tabulated similar designs based on the wider class of generalized cyclic designs, although their designs are no longer adjusted orthogonal. These methods provide a flexible easily constructed class of row–column designs, but have the disadvantage that many of the designs are relatively inefficient since their row components are inefficient. The following example illustrates the method of construction, and shows why the design is not particularly efficient.

Example 5.8
Consider the three-replicate α-design for $v = 12$ treatments in blocks of $k = 4$ given in Example 4.2. The rows within each replicate are reordered to give the following row–column design with $k = 4$ rows and $s = 9$ columns:

0	1	2	4	5	3	7	8	6
3	4	5	8	6	7	10	11	9
6	7	8	9	10	11	0	1	2
9	10	11	0	1	2	5	3	4

It can be verified that the design is adjusted orthogonal. In this case the rows within replicates were reordered to ensure that the row component was (M,S)-optimal. However, it can be seen that treatments $3j$, $3j+1$ and $3j+2$ occur together in three of the four rows ($j = 0,1,2,3$). Hence, the row component has a relatively large number of circuits of length 3, which has been shown in Section 2.5 to be undesirable.

Unless adjusted orthogonality is required, designs which are at least as efficient as, and often more efficient than, those given by John and Eccleston (1986) can be constructed. The same applies to the designs of Lewis and Dean (1991).

Fast computer algorithms capable of producing optimal or near-optimal row–column designs are now available. The early algorithms of Jones (1979), Eccleston and Jones (1980) and Russell, Eccleston and Knudsen (1981) all had the drawback that they could become trapped at local optima. In an attempt to overcome this problem, algorithms using two different approaches have been developed.

Nguyen and Williams (1993b) describe an iterative improvement algorithm which carries out treatment interchanges within the columns of an optimal or near-optimal column component starting design. Treatments are randomized within the columns of the starting design and then interchanges made that have the greatest effect on the (M,S)-optimality objective function trace($\mathbf{W}^2$), where

$$\mathbf{W} = (1/rs)\mathbf{N}_k\mathbf{N}'_k + (1/rk)\mathbf{N}_s\mathbf{N}'_s \qquad (5.18)$$

Note that for equal replication $\mathbf{W}$ is the substantive part of $\mathbf{A}^*$ given in (5.5), and its role is the same as that taken by $\mathbf{N}\mathbf{N}'$ in (2.16) for block designs. The algorithm is repeated many times in an attempt to avoid local optima, and the design with the highest efficiency factor E chosen. The initial column component design for this two-stage process can be obtained using an optimal or near-optimal block design, such as a balanced incomplete block design (Section 1.2) or cyclic design (Section 3.4). Alternatively, a computer algorithm, such as the cyclic design algorithm of Nguyen (1994), can be used to generate an efficient block design.

The second approach, used by Venables and Eccleston (1993), is simulated annealing. Treatment interchanges are made at random, and accepted if they improve the value of a suitably chosen objective function. Local optima are avoided if interchanges that fail to improve the objective function are accepted with some low probability. An appropriate starting design, the objective function, and the acceptance probability function have to be carefully chosen.

These algorithms produce very efficient designs and are capable of handling large values of the design parameters. They can be extended to provide a powerful and flexible method of constructing designs for a wide range of blocking structures, and are discussed in more detail in Section 6.4.

The upper bounds on the average efficiency factor given in

Section 5.5 can again be used to help the user to decide when to stop an algorithm, and to provide measures against which to assess available designs. For larger designs the two-stage approach of Nguyen and Williams (1993b) can produce efficient designs more quickly than simultaneous improvement of rows and columns.

Example 5.9
The two algorithms above produced the following row–column design for $v = 12$ treatments in $k = 4$ rows and $s = 9$ columns:

3	7	4	8	10	2	1	0	6
11	10	1	2	4	9	3	7	5
6	9	8	3	5	4	0	11	7
10	8	11	5	0	6	9	2	1

This design has an average efficiency factor of $E = 0.783$, compared with $E = 0.748$ for the design of Example 5.8. The upper bound is $U = 0.783$.

Example 5.10
Both Venables and Eccleston (1993) and Nguyen and Williams (1993b) discuss the row–column design for $v = 10$ treatments in $k = 5$ rows and $s = 6$ columns. The algorithms quickly produce the following optimal design with $E = 0.815$:

1	3	2	9	4	0
7	6	5	3	0	9
6	0	4	8	7	2
2	1	3	7	5	8
9	8	6	4	1	5

Example 5.11
A strength of the currently available design generation algorithms is the size of the problems that can be handled. A row–column design for $v = 80$ in $k = 12$ rows and $s = 20$ columns with $E = 0.863$, compared with an upper bound of 0.865, can be obtained using the algorithms of Venables and Eccleston (1993) and Nguyen and Williams (1993b). An optimal column component design for $v = 80, k = 12$, given by the dual of the cyclic design with initial blocks (0 1 6), (0 1 9), (0 2 6) and (0 7 10), was used in the two-stage approach of Nguyen and Williams (1993b); this design was generated using the algorithm of Nguyen (1994).

5.9 Nested row–column designs

The row–column designs considered so far have involved ks plots arranged in k rows and s columns. More generally, an experiment may involve g (> 1) groups of row–column designs each with ks plots. These groups may represent a further blocking factor or provide repeats of the basic experimental design. For example, the agricultural experiment of Section 5.1 may be carried out at a number of different locations, so that locations will be a further blocking factor corresponding to the groups in the design. Such designs are called *nested row–column* designs. In addition to the row and column component designs D_k and D_s, there will be a new component design, D_g, corresponding to the groups.

A fully orthogonal nested row–column design is given by taking g repeats of a Latin square design. When $v = k = s$ is small it will be necessary, for instance, to use such a design in order to have sufficient degrees of freedom in the analysis of variance to provide an estimate of experimental error. Variance-balanced designs are obtained by taking repeats of Youden square designs.

Although other row–column designs can also be repeated to provide nested designs this procedure will not, in general, produce the most efficient design. For instance, consider the nested design with $g = 2$ groups given by repeating the design in Example 5.7. While the design is still row-orthogonal, the column component design will now have some pairs of treatments occurring together in four columns and others in two. A better design would be obtained if these column concurrences were 3 and 4.

In practice the most useful type of nested row–column design is the *resolvable row–column* design, given when $v = ks$ and the groups constitute replicates of the treatments. These designs will be discussed in detail in Chapter 6.

Nested designs with $v > ks$ have also received considerable attention, and will now be considered. The groups will be incomplete in that not all treatments will occur in any group, so that D_g is an incomplete block design. For instance, a design with $v = 8$, $g = 4$, $k = 2$ and $s = 3$ is given by

$$
\begin{array}{ccc\quad ccc\quad ccc\quad ccc}
2 & 6 & 3 & \quad 1 & 5 & 2 & \quad 3 & 7 & 4 & \quad 0 & 4 & 1 \\
4 & 7 & 0 & \quad 3 & 6 & 7 & \quad 5 & 0 & 1 & \quad 2 & 5 & 6
\end{array}
$$

Variance-balanced nested row–column designs with binary incomplete blocks rather than complete blocks as groups, have been given by Preece (1967), Singh and Dey (1979), Street (1981), Agrawal and Prasad (1982a), Ipinyomi and John (1985), Cheng

(1986), Sreenath (1989) and Uddin and Morgan (1990, 1991). The information matrix of these designs is again given by (2.11) but with

$$E = \frac{v(k-1)(s-1)}{ks(v-1)} \qquad (5.19)$$

One method of construction, given by Singh and Dey (1979), is to take a balanced incomplete block design for v treatments in b blocks of $k = s^2$ treatments per block and then to set out the s^2 treatments in each of these blocks in a lattice square design of $s + 1$ replicates. However, the resulting balanced row–column design will invariably require a very large number of groups, namely $g = b(s + 1)$.

Other balanced designs can be obtained by cyclically developing one or more initial row–column designs. Agrawal and Prasad (1982a) give four series of designs using this method, which require that $v = tks + 1$ for some $t \geq 1$ and that v is a prime power. Ipinyomi and John (1985) have considered the general problem of obtaining nested row–column designs based on cyclic and generalized cyclic methods of construction and have tabulated a number of designs for $v \leq 15$, nine of which are balanced.

As was the case with incomplete block designs, variance-balanced nested row–column designs will only exist for a limited number of parameter combinations. Most designs involve a large number of groups and have, therefore, limited practical use. Agrawal and Prasad (1982b) give two series of designs, called group divisible and rectangular nested row–column designs respectively, for which the group, row and column component designs are all balanced incomplete block designs. Again these designs, when they do exist, involve a relatively large number of groups.

A flexible family of nested row–column designs can be obtained using cyclical methods of construction. Jarrett and Hall (1982) suggest the use of the generalized cyclic designs of Section 3.6, while Ipinyomi and John (1985) have provided a table of efficient designs for $5 \leq v \leq 15, k \leq 3, s \leq 7$ and $g \leq v$.

Example 5.12
Consider the following design for seven treatments in seven groups of four, with each group set out in two rows and two columns:

```
0 2   1 3   2 4   3 5   4 6   5 0   6 1
4 1   5 2   6 3   0 4   1 5   2 6   3 0
```

The groups of 2×2 squares have been generated by a cyclical development of the first group.

If the three component designs D_k, D_s and D_g are cyclic designs, as they are in Example 5.12, then their information matrices $\mathbf{A}_k$, $\mathbf{A}_s$ and $\mathbf{A}_g$ respectively will be circulant matrices. The canonical efficiency factors of these cyclic nested row–column designs can be obtained directly from the information matrix $\mathbf{A}$ which can be written in the general form

$$\mathbf{A} = \mathbf{A}_k + \mathbf{A}_s - \mathbf{A}_g \tag{5.20}$$

or by calculating

$$e_u = (1/r) \sum_{h=0}^{v-1} (a_{kh} + a_{sh} - a_{gh}) \cos(2\pi hu/v) \tag{5.21}$$

$$(u = 1, 2, \ldots, v - 1)$$

where a_{kh}, a_{sh} and a_{gh} are the elements in the first row and $(h + 1)$th column of the information matrices $\mathbf{A}_k$, $\mathbf{A}_s$ and $\mathbf{A}_g$ respectively. Formula (5.21) follows directly from (3.2) and (5.20).

Each initial group of D_g will usually generate a full arrangement of $g = v$ groups, although partial arrangements with $g < v$ can also be obtained if v and k have a common divisor; see Section 3.2. In order to be able to use (5.21) for designs with $g < v$ it is necessary, however, that not only should D_g be a partial set but that both D_k and D_s should still be cyclic designs, consisting of full sets, partial sets or multiples of partial sets. For example, with $v = 12$ the initial group (0 1 2 4 5 6 8 9 10) will generate a partial cyclic set with $g = 4$ and $r = 3$. Consider the following two ways in which this initial group can be set out in an array with $k = 3$ rows and $s = 3$ columns:

$$
\begin{array}{ccc@{\qquad}ccc}
0 & 1 & 2 & 0 & 1 & 2 \\
5 & 6 & 4 & 5 & 4 & 6 \\
10 & 8 & 9 & 8 & 9 & 10
\end{array}
$$

Cyclic development of the first array will produce a cyclic row–column design for $g = 4$ groups whereas the second array will not; it can be checked that with four groups the column component D_s is not a cyclic design. The second array will, of course, produce a cyclic row–column design with $g = 12$ groups. Alternatively, when $g < v$ generalized cyclic methods of construction can be used.

Example 5.13

The component designs of the following nested row–column design for $v = 8$, $g = 4$, $r = 3$, $k = 2$ and $s = 3$ are all generalized cyclic designs with increment number $m = 2$:

0	1	4		2	3	6		4	5	0		6	7	2
2	5	3		4	7	5		6	1	7		0	3	1

When $g < v$, Ipinyomi and John (1985) show that for some values of v, k, s and r the best cyclic row–column designs are based on partial cyclic sets ($m = 1$), but for other parameter values they are obtained from generalized cyclic designs ($m > 1$).

CHAPTER 6

Resolvable row–column designs

6.1 Introduction

The importance and value of setting out the experimental material into complete replicates has been discussed in Section 4.1 in relation to the resolvability of incomplete block designs. These comments apply also to row–column designs. Suppose the ks plots in each replicate are arranged in k rows and s columns. Then a resolvable row–column design will be needed if it is required to allow for possible differences between both rows and columns within replicates. For example, Williams and Matheson (1994) consider species trials in forestry where plots are usually square and contain a 5×5 or larger grid of trees. In this type of situation a two-dimensional blocking structure of rows and columns within replicates is appropriate, and likely to be much more effective than the one-dimensional blocking structure discussed in Chapter 4.

Resolvable row–column designs are an important special case of nested row–column designs given in Section 5.9; now the groups constitute complete replicates of the treatments, so that $g = r$. For example, a (lattice square) design for $v = 9$ treatments in $r = 2$ replicates each containing $k = 3$ rows and $s = 3$ columns is given by

$$
\begin{array}{ccc\,ccc}
0 & 1 & 2 & 0 & 4 & 8 \\
3 & 4 & 5 & 5 & 6 & 1 \\
6 & 7 & 8 & 7 & 2 & 3
\end{array}
$$

The component designs D_k and D_s correspond to resolvable block designs with rows and columns within replicates respectively as blocks. The component design D_g for replicates is simply a complete block design. The information matrix $\mathbf{A}$ in (5.20) then simplifies to give

$$
\mathbf{A} = r\mathbf{I} - (1/s)\mathbf{N}_k\mathbf{N}'_k - (1/k)\mathbf{N}_s\mathbf{N}'_s + r\mathbf{K} \qquad (6.1)
$$

where $\mathbf{N}_k$ and $\mathbf{N}_s$ are the incidence matrices of the designs D_k

Table 6.1. *Partition of degrees of freedom in a resolvable row–column design*

Stratum	d.f.
Between replicates	$r-1$
Between rows (within replicates)	$r(k-1)$
Between columns (within replicates)	$r(s-1)$
Within rows and columns (within replicates)	$r(k-1)(s-1)$
Total	$rks-1$

and D_s respectively. The partition of the degrees of freedom in the analysis of variance of a resolvable row–column design is given in Table 6.1, and can be seen to involve four blocking strata. The extension of these designs to allow for the case where the r replicates are contiguous is considered in Section 6.5.

6.2 Lattice square designs

Lattice square designs were introduced by Yates (1940a). They are resolvable row–column designs for $v = s^2$ treatments set out in r replicates with $k = s$ rows and columns. In Section 4.2, balanced lattice designs were defined for $v = s^2$ treatments using complete sets of mutually orthogonal Latin squares. The s^2 treatments were set out in an $s \times s$ array and $s+1$ replicates of s blocks each obtained by superimposing rows, columns and each of the $s - 1$ orthogonal Latin squares onto the array. Such designs exist for any s which is a prime or a power of a prime. In a balanced lattice square design the rows and columns correspond to these replicates. They are arranged in such a way that for s odd half of the replicates constitute the rows and the other half the columns, whereas for s even each replicate is used both for rows and for columns. Hence a balanced lattice square design requires $r = \frac{1}{2}(s + 1)$ replicates for s odd and $r = s + 1$ replicates for s even.

The lattice square design for $s = 3$ given in Section 6.1 has the rows of the two squares corresponding to two of the $s + 1 = 4$ replicates of a balanced lattice design while the columns correspond to the other two replicates. Since D_k and D_s together give a balanced lattice design, it follows from (6.1) that this lattice square design is variance-balanced. An example of a balanced lattice square design for s even is as follows:

Example 6.1
For $s = 4$ the five replicates are

		Replicate		
1	2	3	4	5
0 1 2 3	0 4 8 12	0 5 10 15	0 6 11 13	0 7 9 14
4 5 6 7	5 1 13 9	11 14 1 4	7 1 12 10	1 6 8 15
8 9 10 11	10 14 2 6	13 8 7 2	9 15 2 4	2 5 11 12
12 13 14 15	15 11 7 3	6 3 12 9	14 8 5 3	3 4 10 13

Note that the columns of the first replicate correspond to the rows of the second replicate, the columns of the second to the rows of the third, and so on. Hence, both the row and column component designs D_k and D_s are balanced lattice designs. It then follows from (6.1) that the lattice square design with $v = 16$, $r = 5$ and $k = s = 4$ is a variance-balanced resolvable row–column design.

Lattice square designs can be defined for r other than $\frac{1}{2}(s + 1)$ and $s + 1$, but such designs will no longer be variance-balanced. Cochran and Cox (1957) consider designs for $r < \frac{1}{2}(s + 1)$. For s odd it is just a matter of choosing any r replicates from the full set of $\frac{1}{2}(s + 1)$ replicates. Pairs of treatments will concur in rows or columns either once or not at all. For s even, however, more care is needed to ensure that row–column treatment concurrences are not greater than one. For example, by choosing replicates 1 and 3 in Example 6.1, a two-replicate lattice square design is obtained such that pairs of treatments concur at most once in rows or columns. Choosing replicates 1 and 2, however, leads to a lattice square design in which treatments 0, 4, 8 and 12, for instance, all concur twice, namely in column 1 of replicate 1 and row 1 of replicate 2. It follows from the discussion in Section 2.5 on (M,S)-optimality that it is better to choose a design which minimizes the sum of squares of treatment concurrences. Hence using replicates 1 and 3 of the design of Example 6.1 will give a more efficient two-replicate lattice square design than replicates 1 and 2.

With $\frac{1}{2}(s+1) < r < s+1$, some pairs of treatments must concur more than once in rows and columns. Federer (1955) considered lattice square designs in this range and restricted attention to the situation where rows of one replicate are duplicated as columns in another replicate.

Example 6.2

For $s = 4$ and $r = 3$, the lattice square design

Replicate												
1				2				3			
0	1	2	3	0	4	8	12	0	5	10	15
4	5	6	7	5	1	13	9	11	14	1	4
8	9	10	11	10	14	2	6	13	8	7	2
12	13	14	15	15	11	7	3	6	3	12	9

has the columns of replicate 1 duplicated as the rows of replicate 2 and the columns of replicate 2 duplicated as the rows of replicate 3; i.e. there are two sets of row–column duplications. Note that the design comprises the first three replicates of Example 6.1, and that the component designs D_k and D_s are both triple lattice designs; see Section 4.2.

The number d of row–column duplications can vary. For example, by replacing the third replicate in Example 6.2 by

0	5	10	15
1	4	11	14
2	7	8	13
3	6	9	12

a lattice square design with $d = 3$ is obtained.

Based on results of Williams, Ratcliff and van Ewijk (1986) and A. M. Kshirsagar (Personal Communication), it can be shown that a generalized inverse $\mathbf{\Omega}$ of the information matrix $\mathbf{A}$ of a lattice square design is given by

$$\mathbf{\Omega} = \frac{1}{r} \left\{ \mathbf{I} + \frac{1}{s(r-1)} (\mathbf{N}_k \mathbf{N}_k' + \mathbf{N}_s \mathbf{N}_s') + \frac{2}{s(r-1)(r-2)} \mathbf{T} \right\} \tag{6.2}$$

where $\mathbf{T}$ is a symmetric matrix whose (i, j)th element is the number of times treatments i and j concur in rows of replicates that are duplicated as columns of other replicates. Thus for Example 6.2, $\mathbf{T}$ would be calculated only from the treatment concurrences in the rows of replicates 2 and 3 (or equivalently the columns of replicates 1 and 2).

Now $\text{trace}(\mathbf{N}_k \mathbf{N}_k') = \text{trace}(\mathbf{N}_s \mathbf{N}_s') = rs^2$ and $\text{trace}(\mathbf{T}) = ds^2$ so that

$$\text{trace}(\mathbf{\Omega}) = \frac{1}{r} \left\{ s^2 + \frac{2rs}{r-1} + \frac{2ds}{(r-1)(r-2)} \right\}$$

Also the eigenvalue corresponding to the unit eigenvector of Ω is

$$\frac{1}{r}\left\{1+\frac{2r}{r-1}+\frac{2d}{(r-1)(r-2)}\right\}$$

Hence, using (2.9), the average efficiency factor of a lattice square design is

$$E = \frac{(s+1)(r-1)(r-2)}{(s+1)(r-1)(r-2)+2\{d+r(r-2)\}} \tag{6.3}$$

The value of E increases as the value of d decreases.

For $r \leq \frac{1}{2}(s+1)$ lattice square designs with $d = 0$ can be constructed. In this case $\mathbf{T} = \mathbf{0}$ and (6.2) reduces to

$$\Omega = \frac{1}{r}\left\{\mathbf{I}+\frac{1}{s(r-1)}(\mathbf{N}_k\mathbf{N}'_k+\mathbf{N}_s\mathbf{N}'_s)\right\}$$

analogous to (4.3) for a square lattice design. Further, (6.3) gives

$$E = \frac{(s+1)(r-1)}{(s+1)(r-1)+2r} \tag{6.4}$$

All lattice square designs with $d = 0$ are adjusted orthogonal in that they satisfy the condition that each row of the design has one treatment in common with each column, i.e. $\mathbf{N}'_k\mathbf{N}_s = \mathbf{J}$. It can then be verified that (6.4) is also given by

$$E = (E_k^{-1}+E_s^{-1}-1)^{-1}$$

where from (4.2),

$$E_k = E_s = \frac{(s+1)(r-1)}{r(s+2)-(s+1)}$$

are the average efficiency factors of the component square lattice designs. Since these components are optimal, as shown by (4.11), it follows from (5.15) that lattice square designs where pairs of treatments concur at most once in rows or columns are optimal.

For the balanced lattice square design with s odd and $r = \frac{1}{2}(s+1)$, all the canonical efficiency factors are equal to the average efficiency factor

$$E = (s-1)/(s+1) \tag{6.5}$$

obtained from (6.4). The balanced lattice square designs for s even are also optimal. They have $d = r = s+1$ and (6.3) reduces to (6.5). They are not, however, adjusted orthogonal as this requires $d = 0$. This can be clearly seen in the design in Example 6.1 which has $d = 5$ and $E = 0.6$.

For the balanced lattice square designs with either $r = \frac{1}{2}(s + 1)$ or $s + 1$, the information matrix $\mathbf{A}$ has the form (2.11) with E given by (6.5).

6.3 Construction using α-designs

Lattice square designs provide efficient resolvable row–column designs when $k = s$, but designs are also required when $k \neq s$. Not many such designs are available in the literature. Some of the designs listed by Preece (1967) can be used; Cheng (1986) defined the class of balanced complete block row and column designs but only gave balanced lattice square designs as examples; and Ipinyomi and John (1985) included 17 resolvable row–column designs in their table of nested designs. Patterson and Robinson (1989) and Bailey and Patterson (1991) gave some two-replicate row–column designs which will be discussed separately in Section 6.6. In practice, however, experimenters need to have a wide range of efficient resolvable row–column designs at their disposal. A similar requirement in block designs led to the development of α-designs. As was the case with α-designs, a combination of theory and computer search provides the means for generating a range of efficient resolvable row–column designs.

One method of constructing resolvable row–column designs is to use the α-designs of Section 4.4. For instance, suppose that each of the three replicates of the α-design in Example 4.2 consisted of a 4×3 array of plots. The design would then provide a resolvable row–column design for $v = 12$, $k = 4$, $s = 3$ and $r = 3$. The column component design D_s is very efficient since it is given by the α-design. The row component design D_k on the other hand is disconnected, with treatment concurrences of either 0 or 3. Consequently, the overall row–column design can be shown to be relatively inefficient. However, by rearranging the order of the treatments in the columns of the design, the row component design and hence the overall row–column design can be improved, while leaving D_s unchanged.

Nguyen and Williams (1993a) have developed a computer algorithm that interchanges treatments within the columns of a resolvable row–column component design D_s with the aim of finding an efficient resolvable row–column design, in the sense of minimizing the objective function trace($\mathbf{W}^2$) where $\mathbf{W}$ is given by (5.18); this function is the same as that used in Section 5.8. Using this algorithm on the α-design of Example 4.2 produced the following efficient resolvable row–column design:

			Replicate					
	1			2			3	
3	7	5	8	6	7	10	8	2
9	10	11	4	1	11	5	3	4
0	1	2	9	5	2	0	11	6
6	4	8	0	10	3	7	1	9

It can be seen that the row component design D_k now has treatment concurrences of just 0 and 1.

Example 6.3

For $v = 12$ treatments in $r = 4$ replicates each with $k = 3$ rows and $s = 4$ columns, an (optimal) rectangular lattice design can be chosen for the column component design D_s. Then using the Nguyen and Williams (1993a) algorithm the design

							Replicate								
	1				2				3				4		
4	1	2	3	6	0	3	2	7	3	0	1	5	2	1	0
8	5	6	7	11	7	4	5	9	6	5	4	10	4	6	7
0	9	10	11	1	10	9	8	2	8	11	10	3	11	8	9

is obtained with $E = 0.535$ compared with an upper bound of 0.537.

The algorithm is particularly suitable for the construction of resolvable row–column designs with many treatments. The design generation package ALPHA+ of Williams and Talbot (1993) uses this algorithm to construct efficient resolvable row–column designs. First the package generates a good α-design, and then uses this design as the column component design D_s in the algorithm of Nguyen and Williams (1993a). The average efficiency factor of the designs generated are compared with upper bounds obtained using the results of Section 6.7.

Example 6.4

ALPHA+ can be used to construct a resolvable row–column design for $v = 45$ treatments in $r = 4$ replicates each with $k = 9$ rows and $s = 5$ columns. The optimal α-array

$$
\begin{array}{cccc}
0 & 0 & 0 & 0 \\
0 & 1 & 2 & 4 \\
0 & 2 & 4 & 3 \\
0 & 3 & 1 & 2 \\
0 & 4 & 3 & 1 \\
0 & 0 & 1 & 1 \\
0 & 1 & 3 & 0 \\
0 & 2 & 0 & 4 \\
0 & 3 & 2 & 3
\end{array}
$$

is first generated for a resolvable incomplete block design with nine plots per block. Treatments are then interchanged within the columns of the α-design to give the resolvable row–column design

Replicate			
1	2	3	4
41 7 3 14 35	13 33 40 36 37	1 18 29 32 5	44 0 41 42 39
16 32 28 24 10	38 20 16 35 6	27 25 31 39 23	31 32 33 30 9
1 42 43 19 20	25 21 22 42 24	24 2 38 41 40	40 6 7 4 21
6 17 33 29 25	44 27 3 4 5	8 9 10 6 26	1 28 37 8 35
36 27 38 9 30	32 14 34 11 43	15 11 3 13 42	10 2 3 25 43
11 12 8 39 40	1 0 9 23 12	17 28 21 30 14	22 36 11 38 5
21 37 23 44 15	7 39 15 10 18	36 44 12 20 7	14 15 24 12 26
26 22 18 4 0	26 2 28 29 30	34 35 19 22 23	27 19 29 16 13
31 2 13 34 5	19 8 41 17 31	43 37 0 4 16	18 23 20 34 17

with $E = 0.697$ compared with an upper bound of 0.718. The efficiency factors for the component designs D_k and D_s are respectively 0.756 and 0.890 compared with upper bounds of 0.788 and 0.890.

The Nguyen and Williams algorithm interchanges treatments within the columns of a resolvable column component design. An alternative interchange algorithm based on simulated annealing has been given by John and Whitaker (1993); simulated annealing algorithms have already been briefly discussed in Section 5.8. Interchange algorithms provide a powerful means of generating a wide range of good designs, and will now be considered in more detail.

6.4 Interchange algorithms

As a result of advances in computer technology, attention is increasingly being focused on the development of fast computer algorithms to generate designs. These algorithms are capable of producing good designs for a whole range of parameter values, and for a wide range of blocking structures. Such algorithms have already been referred to in Sections 2.4, 3.4, 4.5, 5.8 and 6.3 for the construction of different types of block and row–column designs. All of these algorithms are based on some type of interchange procedure.

An interchange algorithm begins with the choice of a suitable starting design. A pair of treatments is chosen, and the effect on some objective function of interchanging this pair determined. A decision is then taken to either change the design by making the interchange or to leave the design unchanged. The process is repeated with another pair of treatments, and the procedure continued until either an optimal design is found, or a stopping rule satisfied, or the user intervenes. The starting design, the way pairs of treatments are chosen, the interchange decision rule, the objective function and stopping criteria are all aspects that have be considered.

The simplest way to choose a starting design is to fill the plots of the design with the numbers 1 to v repeatedly. For instance, a resolvable row–column design would have the first row filled with treatments $1, 2, \ldots, k$, the next row with $k + 1, k + 2, \ldots, 2k$, and so on. The resulting starting design would clearly be resolvable, but all replicates would be the same and the design, therefore, disconnected. This will not usually matter as the interchange algorithm will quickly change it. However, some variant of this method could be used instead. In the row–column design algorithm of Nguyen and Williams (1993a) the column component design is fixed; this design then becomes the starting design. Repeated running of the algorithm may require different starting designs each time. This can be achieved, for instance, by randomly permuting the treatments in the previous starting design.

The way pairs of treatments are chosen for interchange depends on the blocking structure of the design. For instance, in a block design interchanging two treatments from the same block has no effect on the design. Only pairs of treatments coming from different blocks should, therefore, be considered for interchange. In a resolvable design, in order to preserve resolvability the pairs will come from the same replicate. In a row–column design, if the column component design is fixed pairs of treatments will be restricted to those belonging to the same column.

The choice of which pair to consider for interchange can be made in a number of different ways. Pairs can be chosen systematically from the set of all possible pairs. If an interchange takes place, then the next pair in the sequence can be chosen or the process can be started from the beginning again. Alternatively, pairs can be chosen at random. This is the method used in simulated annealing algorithms, such as those described by Jansen, Douven and van Berkum (1992), John and Whitaker (1993) and Venables and Eccleston (1993). Another strategy is to identify the pair of treatments which would make the greatest improvement to the design if interchanged. Nguyen and Williams (1993a) use a variant of this method whereby a column is chosen at random and then the pair of treatments in this column leading to the greatest improvement identified.

Once a pair of treatments has been chosen, the effect on the chosen objective function of interchanging these treatments is calculated. Based on this effect a decision is then made as to whether these treatments are interchanged in the design. An obvious rule is to make the interchange only if by doing so an improved design results. Unfortunately, this does not guarantee that the best, or even a good, design will be found; the algorithm can get stuck at a local optimum. Repeated application of the algorithm with different starting designs is one way to overcome this difficulty, and is the approach used by Nguyen and Williams (1993a). An alternative is to allow interchanges to take place even when no improvement results; this is a key feature of simulated annealing algorithms. Thus an interchange that would improve the design is always made, but with some small probability an interchange that does not lead to an improvement is also made.

The final aspect of an interchange algorithm is the stopping rule. This can be based on a comparison of the average efficiency factor E with its upper bound U. The algorithm should stop when a design is found with $E = U$. However, in many cases the bound will not be reached. The user should then be able to intervene to stop the program, usually when the value of E is deemed to be close enough to U.

An appropriate objective function is the average efficiency factor E. However, E involves either matrix inverse or eigenvalue calculations. Simpler, and computationally cheaper, functions can be based on the (M,S)-optimality criterion. For block designs this corresponds to minimizing trace$\{(\mathbf{NN}')^2\}$. The average efficiency factor would then be calculated only when a design is found which decreases the value of this objective function.

For resolvable row–column designs the corresponding objective function is $\text{trace}(\mathbf{W}^2)$, where $\mathbf{W}$ is given by (5.18). The final choice of design can still be based on the average efficiency factor. However, other design requirements may have to be considered. If, for instance, the column factor turns out to be unimportant then the design might be analysed as a block design with the rows as the blocking factor. In this case it is important that the row component design constitutes a good block design. Ideally, therefore, it might be desirable to optimize the row and column component designs as well as the row–column design. Since it may be impossible to optimize all three designs simultaneously, some compromise will usually be necessary. The form this compromise takes will depend on the amount of importance that is to be attached to the three designs.

An approach adopted by John and Whitaker (1993) is to use a weighted linear combination of the objective functions of the three designs. Now

$$\text{trace}(\mathbf{W}^2) = \text{trace}(\mathbf{W}_k^2) + \text{trace}(\mathbf{W}_s^2)$$
$$+ (2/r^2 v)\,\text{trace}\{(\mathbf{N}_k'\mathbf{N}_s)'(\mathbf{N}_k'\mathbf{N}_s)\} \qquad (6.6)$$

where

$$\mathbf{W}_k = (1/rs)\mathbf{N}_k\mathbf{N}_k' \qquad \text{and} \qquad \mathbf{W}_s = (1/rk)\mathbf{N}_s\mathbf{N}_s'$$

The first two terms on the right-hand side of (6.6) depend on the row and column component designs respectively, while the third term is minimized when the row–column design is adjusted orthogonal. Let the objective function f be

$$f = \omega_1\,\text{trace}(\mathbf{W}_k^2) + \omega_2\,\text{trace}(\mathbf{W}_s^2) + \omega_3\,\text{trace}(\mathbf{W}^2) \qquad (6.7)$$

where ω_1, ω_2 and ω_3 are weights. A simple choice of weights would be $\omega_1 = \omega_2 = 0$ and $\omega_3 = 1$. However, if it was also important to ensure a good row component design, say, then a value of $\omega_1 > 0$ would be used; an appropriate choice can be determined by trial and error. With f as the objective function, adjusted orthogonal row–column designs can be readily constructed using the weights $\omega_1 = \omega_2 = -1$ and $\omega_3 = 1$. This can be verified from (6.6) and (6.7). A design with these weights may not be particularly efficient though, so that additional weight on the row and column component designs would be desirable.

Example 6.5
The John and Whitaker algorithm with weights $\omega_1 = \omega_2 = -0.7$

and $\omega_3 = 1$ gave the following resolvable row–column design for $v = 12$, $k = 3$, $s = 4$ and $r = 4$

Replicate															
1				2				3				4			
1	4	6	10	1	10	6	4	2	1	11	6	4	0	9	8
0	5	9	2	5	9	0	2	8	0	4	9	3	10	7	5
3	11	7	8	8	11	3	7	10	7	5	3	2	11	1	6

It can be verified that this is an adjusted orthogonal design since each row of the design has one treatment in common with each column. The average efficiency factor is $E = 0.509$, compared to $E = 0.535$ for the design of the same size given in Example 6.3. Although the third term on the right-hand side of (6.6) has been minimized, the row and column component are less efficient than those in the design of Example 6.3. In this case, therefore, a large reduction in the value of E results from the additional constraint of adjusted orthogonality.

6.5 Latinized row–column designs

Latinized block designs were discussed in Section 4.6 and provide a design structure for the situation when replicates are contiguous. Very often, however, the plots in a contiguous replicate design will form a rectangular array, in which case it will be sensible to allow for possible differences between rows within replicates in the design. Such designs are called *Latinized row–column* designs. For instance, in the cotton variety trial layout in Figure 4.1, there could be a yield trend as one moves away from the irrigation channel. The replicates accommodate this variation to some extent in a Latinized block design, but greater gains in precision might be achieved by using a design that allows for possible between-row variation.

A feature of modern experimental design is the facility to include, if appropriate, extra blocking structures in the design. This ability is due to a large extent to the ready availability of computer software both for the construction of designs and for the analysis of general experimental layouts. With the layout in Figure 4.1, a Latinized row–column design incorporates all the appropriate blocking structures.

An additional blocking stratum will now need to be added to those discussed in Section 4.6 to represent the variation between rows (within replicates). The full partition of degrees of freedom

Table 6.2. *Partition of degrees of freedom in a Latinized row–column design*

Stratum	d.f.
Between replicates	$r-1$
Between long columns	$s-1$
Between rows (within replicates)	$r(k-1)$
Between columns (within replicates)	$(r-1)(s-1)$
Within rows and columns (within replicates)	$r(k-1)(s-1)$
Total	$rks-1$

in a Latinized row–column design is given in Table 6.2.

To illustrate the issues involved in constructing a Latinized row–column design, consider the row–column α-design given in Example 5.8. Switching rows and columns, the layout has the structure that can be considered suitable for a 9×4 arrangement of plots, i.e.

Replicate

1	0	3	6	9
	1	4	7	10
	2	5	8	11
2	4	8	9	0
	5	6	10	1
	3	7	11	2
3	7	10	0	5
	8	11	1	3
	6	9	2	4

The above Latinized design has no treatment occurring more than once in any long column. The design is adjusted orthogonal but the columns within-replicates design D_s is disconnected, so that the average efficiency factor of the resolvable row–column design will be low. On the other hand by taking three of the replicates of the Latinized rectangular lattice design given in Section 4.6, another layout for a 9×4 arrangement of plots is obtained. In this case, however, the rows within-replicates design D_k is disconnected.

A better design for this layout would be an optimal or near-optimal resolvable row–column design that also has the Latinized property that treatments do not appear more than once in long columns. The design

Replicate				
1	0	1	2	3
	8	5	10	7
	4	9	6	11
2	6	0	3	5
	1	7	4	2
	11	10	9	8
3	9	8	0	1
	7	6	5	4
	2	3	11	10

is a Latinized row–column design with average efficiency factors $E_k = 0.680$ and $E_s = 0.759$ for D_k and D_s respectively, and $E = 0.508$ for the row–column design. Note that D_k is a rectangular lattice design and so is optimal, while the average efficiency factor for D_s is only slightly less than the corresponding value for the above adjusted orthogonal design, namely 0.767. Hence for a 9×4 array of plots, the Latinized row–column design having treatments efficient relative to long columns as well as rows and columns within replicates is to be preferred.

Again computer algorithms permit the construction of efficient Latinized resolvable row–column designs for a wide range of parameter values. Starting with an efficient Latinized α-design, the algorithm of Nguyen and Williams (1993a) can then be used to interchange treatments in columns within replicates to produce an efficient resolvable row–column design. As already stated in Section 6.3, this algorithm has been incorporated into the design package ALPHA+ of Williams and Talbot (1993).

Example 6.6
An experimental design was required to test the juvenile performance of 40 provenances of *Eucalyptus camaldulensis* in a glasshouse. There tends to be considerable growth variation in glasshouses, both from one end to the other and from the sides into the middle. Hence a Latinized row–column design is particularly suitable for this environment. A 30×8 array of plots was available in the glasshouse and so a design for $k = 5$, $s = 8$ and $r = 6$ was generated by ALPHA+ with $E_s = 0.808$ for the first-stage

Latinized α-design and $E = 0.707$ for the row–column design. The design is as follows:

Replicate

1	17	18	19	28	5	22	15	0
	9	26	35	12	37	38	31	32
	33	10	3	20	13	30	7	24
	1	34	11	36	21	14	39	16
	25	2	27	4	29	6	23	8
2	21	31	4	15	17	27	8	38
	30	0	14	34	26	7	37	29
	39	13	23	24	6	18	19	11
	2	3	32	5	16	9	28	1
	12	22	33	25	35	36	10	20
3	28	16	20	31	34	23	13	17
	3	4	5	21	11	12	26	27
	18	29	9	6	22	35	24	14
	38	39	0	10	32	25	36	2
	15	19	30	33	7	8	1	37
4	35	11	31	2	20	15	16	9
	29	8	18	32	39	21	33	28
	24	36	37	38	3	26	5	23
	10	30	1	19	25	4	22	34
	7	17	12	13	14	0	27	6
5	31	32	25	26	27	28	29	30
	19	20	21	22	23	24	17	18
	36	37	38	39	0	33	34	35
	5	6	7	8	1	2	3	4
	14	15	16	9	10	11	12	13
6	34	21	36	3	9	32	25	19
	27	1	15	30	24	39	6	12
	20	35	2	23	38	5	0	33
	8	14	29	16	31	10	18	26
	13	28	22	37	4	17	11	7

Alternatively, the simulated annealing algorithm of John and Whitaker (1993) can be used. Consider the objective function f

given in (6.7). Suppose a further term

$$\omega_4 \operatorname{trace}(\mathbf{N}_c \mathbf{N}_c') \tag{6.8}$$

is added to this function, where $\mathbf{N}_c \mathbf{N}_c'$ is the concurrence matrix of the long column design and ω_4 is the weight. For a given ω_4, this term is minimized when the long column design is binary or M-optimal, which is the required Latinized property. Choosing ω_4 to be large relative to the other weights forces the algorithm to generate efficient Latinized resolvable row–column designs.

The use of weighted linear combinations of objective functions can be extended to allow designs with other blocking structures to be constructed. Suppose, for instance, that the replicates are to be set out in a rectangular array, and that it is desired to eliminate the effects of both long columns and long rows. That is, the resolvable row–column design is Latinized in both directions. Then to the functions (6.7) and (6.8) another term

$$\omega_5 \operatorname{trace}(\mathbf{N}_r \mathbf{N}_r')$$

is added, where $\mathbf{N}_r \mathbf{N}_r'$ is the concurrence matrix of the long row design and ω_5 the weight.

Example 6.7
Suppose the replicates of a resolvable row–column design with $v = 12$, $k = 3$, $s = 4$ and $r = 4$ are set out in a 2×2 array. With weights $\omega_1 = \omega_2 = 0$ and $\omega_3 = \omega_4 = \omega_5 = 1$ the algorithm of John and Whitaker (1993) produced the design

Replicate 1					Replicate 3			
5	6	2	3		10	9	11	0
4	1	11	10		7	6	2	8
0	7	9	8		1	4	3	5
Replicate 2					Replicate 4			
3	4	0	7		8	1	5	9
9	2	10	5		11	0	6	3
1	8	6	11		4	2	10	7

with $E = 0.532$ compared with an upper bound of 0.537. Note that no treatment occurs more than once in any long column or long row, i.e. both long column and long row designs are binary.

The additional blocking factors impose constraints that can result in some loss in the average efficiency factor of the resolvable row–column design. A similar point was made in Section 4.6.2 when it was shown that if $k \geq s$ a Latinized block design cannot be

optimal. With replicates contiguous in one direction only, the need
to keep the rectangular array as compact as possible usually means
in practice that $k < s$. In this situation no loss in the average
efficiency factor will occur.

It should be noted that the optimality criterion used for a Latinized row–column design is just the average efficiency factor of
the resolvable row–column design. The average efficiency factor of
the design using long columns (or rows) as blocks is not taken into
account; there is just the requirement that the long column (or
row) design be binary.

6.6 Two-replicate row–column designs

Resolvable two-replicate block designs for $v = ks$ treatments with
k plots per block were discussed in Section 4.7 and where it
was shown that their construction could conveniently be obtained
from a symmetric block design, called the contraction, with s
treatments. This idea has been extended by Bailey and Patterson
(1991) to the construction of resolvable two-replicate row–column
designs. Consider the following two-replicate design for $v = 28$
treatments, $k = 4$ rows and $s = 7$ columns:

Replicate		0							1					
Column		0	1	2	3	4	5	6	0	1	2	3	4	5	6
Row	0	0	1	2	3	4	5	6	0	22	18	19	6	9	24
	1	7	8	9	10	11	12	13	8	17	23	21	11	5	20
	2	14	15	16	17	18	19	20	16	12	3	4	15	27	7
	3	21	22	23	24	25	26	27	25	2	13	10	26	14	1

Following the method used in Section 4.7, consider the four
treatments in column 0 of replicate 1. Treatment 0 occurs in row
0 and column 0 of replicate 0, treatment 8 in row 1 column 1,
treatment 16 in row 2 column 2, and treatment 25 in row 3 column
4. Proceeding in this manner with the treatments in the other
columns of replicate 1 leads to the row–column contraction

$$
\begin{array}{ccccccc}
00 & 31 & 24 & 25 & 06 & 12 & 33 \\
11 & 23 & 32 & 30 & 14 & 05 & 26 \\
22 & 15 & 03 & 04 & 21 & 36 & 10 \\
34 & 02 & 16 & 13 & 35 & 20 & 01
\end{array}
$$

This contraction consists of two treatment factors at four and seven
levels respectively, the levels referring to the rows and columns
respectively of replicate 0. The rows given by the first factor is the

contraction of the row component design D_k with four treatments in blocks of seven, while the columns given by the second factor is the contraction of the column component design D_s with seven treatments in blocks of four. The row–column contraction is in fact a single replicate factorial row–column design (see Section 9.8).

The properties of the row–column contraction are linked to those of the two-replicate row–column design, but the connection is more complicated than was the case in Section 4.7. If in the row–column contraction, however, the first factor is orthogonal to columns and the second factor is orthogonal to rows, then it follows that the two-replicate row–column design will be adjusted orthogonal. Then from (4.8) and (5.13) the average efficiency factor of the two-replicate design can be written as

$$E = \frac{v-1}{v-1-2(s-1)-2(k-1)+4(s-1)E_s^{*-1}+4(k-1)E_k^{*-1}}$$

(6.9)

where E_k^* and E_s^* are the average efficiency factors for the row and column contractions respectively.

For some parameter values it is possible to demonstrate the optimality of the row–column contraction and hence of the associated two-replicate row–column design. For example, in the above 4×7 row–column contraction, the symmetric design for the second factor is a 4×7 Youden square (Section 5.7) and so is optimal. In addition, the columns of the first factor give a complete block design while the rows give a non-binary design with incidence matrix $\mathbf{J} + \mathbf{N}$, where $\mathbf{N}$ is the incidence matrix of a balanced incomplete block design. Thus it follows from Section 2.5 that the row–column design for the first factor is also optimal. Furthermore, because the first factor is orthogonal to columns and the second factor is orthogonal to rows, the derived two-replicate row–column design is adjusted orthogonal and so from (5.13) it is an optimal 4×7 row–column design for 28 treatments. In particular $E_k^* = 48/49$, $E_s^* = 7/8$ so that (6.9) becomes $E = 0.511$.

The above single replicate factorial row–column design for two factors at four and seven levels respectively is an example of a *double Youden rectangle*, a precise definition of which has been given by Bailey (1989). When the contraction is a double Youden rectangle then the average efficiency factors E_k^* and E_s^* are given by

$$E_k^* = 1 - \frac{s'(k - s')}{s^2(k - 1)} \quad \text{and} \quad E_s^* = 1 - \frac{k'(s - k')}{k^2(s - 1)}$$

where $s' = s$ modulo k and $k' = k$ modulo s; see (2.34).

Double Youden rectangles have been studied in detail by Preece; see Preece (1994) for references. However, they are known to exist for very few combinations of k and s. For $v \leq 100$ double Youden rectangles are currently only available for $k = s - 1$ with $s > 3$, and for the 4×7, 4×13, 5×11 and 6×11 cases. From a practical point of view, optimal or near-optimal row–column contractions are required for other values of k and s. Interchange algorithms again provide a method of generating designs for a wide range of parameter values. Williams and John (1995) have described an algorithm for the construction of factorial row–column designs; see Section 9.8. It can generate optimal or near-optimal single replicate two-factor row–column designs which can be used as row–column contractions to construct efficient resolvable two-replicate row–column designs.

Example 6.8

A resolvable two-replicate design for $v = 50$, $k = 5$ and $s = 10$ can be constructed by first generating a near-optimal single replicate row–column design for two factors at five and ten levels. The Williams and John (1995) algorithm generates the following row–column contraction:

23	45	04	37	09	40	32	18	21	16
31	28	27	06	43	12	49	00	15	34
47	02	35	20	14	38	11	46	03	29
19	33	10	44	22	01	07	25	36	48
08	17	41	13	30	24	26	39	42	05

This contraction is column-orthogonal for the first factor and row-orthogonal for the second. Hence (6.9) can be used with $E_k^* = 1.000$ and $E_s^* = 0.885$, to give an average efficiency factor of $E = 0.615$ for the derived two-replicate row–column design.

6.7 Upper bounds for the average efficiency factor

A simple upper bound for the average efficiency factor of a resolvable row–column design is given by the arithmetic mean of the canonical efficiency factors e_i $(i = 1, 2, \ldots, v - 1)$. In this case it is

$$\bar{e} = \frac{v - k - s + 1}{v - 1} \qquad (6.10)$$

Another bound is provided by (5.16), where U_k and U_s are upper bounds for the component resolvable designs given in Section 4.10. This bound will be denoted by U_0. Note that U_0 can be used for resolvable row–column designs because the information matrix $\mathbf{A}$ in (6.1) has the same form as (5.3) for a general row–column design. When $k = s$ and $r \leq \frac{1}{2}(s + 1)$ U_0 is given by (6.4). For a Latinized row–column design with $k > s$, an upper bound for the efficiency factor of a Latinized block design would be used for U_s; again such a bound is given in Section 4.10.

John and Street (1992) provide an alternative upper bound based on U_1 in (2.27). A lower bound for the corrected second moment $S_2 = \sum(e_i - \bar{e})^2$ is required for U_1. Now

$$\begin{aligned} \sum e_i^2 &= (1/r^2)\,\mathrm{trace}(\mathbf{A}^2) \\ &= (v - 1) - 2(k + s) + \mathrm{trace}(\mathbf{W}^2) \end{aligned}$$

where $\mathbf{W}$ is given by (5.18).

Let λ_{ki} and λ_{sj} be the distinct treatment concurrences in the component designs D_k and D_s respectively. Let $\mu_{ij} = (rs)^{-1}\lambda_{ki} + (rk)^{-1}\lambda_{sj}$ and let n_{ij} be the number of times μ_{ij} occurs in $\mathbf{W}$. Then

$$\mathrm{trace}(\mathbf{W}^2) = \frac{(k + s)^2}{v} + \sum_i \sum_j n_{ij}\mu_{ij}^2 \qquad (6.11)$$

If n_{ki} and n_{sj} are respectively the number of times λ_{ki} and λ_{sj} occur in the concurrence matrices $\mathbf{N}_k\mathbf{N}_k'$ and $\mathbf{N}_s\mathbf{N}_s'$, then the values of n_{ij} must satisfy, for all i and j,

$$\sum_j n_{ij} = n_{ki}, \qquad \sum_i n_{ij} = n_{sj} \qquad (6.12)$$

Thus to obtain a lower bound for S_2 it is necessary to minimize (6.11) subject to (6.12).

Based on the desirability of having good component designs, John and Street (1992) assume that the range of the off-diagonal elements in the concurrence matrices of the component designs D_s and D_k is at a minimum; see Section 4.5. Assuming without loss of generality that $k \geq s$, this means that the treatment concurrences in D_k should differ by at most one and those in D_s by at most two. For example, consider the row–column design for $v = 12$, $k = 4$ and $s = 3$ given in Section 6.3. D_k has treatment concurrences of zero and one but, because $k > s$, D_s has treatment concurrences of zero, one and two.

Thus for D_k the treatment concurrences are $\lambda_{k1} < \lambda_{k2}$ with $\lambda_{k2} - \lambda_{k1} = 1$, and the only possible non-zero n_{ki} are n_{k1} and n_{k2}. Similarly, for D_s the concurrences are $\lambda_{s1} < \lambda_{s2} < \lambda_{s3}$ with $\lambda_{s3} - \lambda_{s1} = 2$, and the only possible non-zero n_{ij} are n_{s1}, n_{s2} and n_{s3}. Now define

$$\lambda_k = \frac{r(s-1)}{v-1}, \qquad \lambda_s = \frac{r(k-1)}{v-1}$$

and let α_k and α_s be the fractional parts of λ_k and λ_s respectively. Further λ_{k1} is the integer part of λ_k, i.e. $\lambda_{k1} = [\lambda_k]$. If $\alpha_s > 1 - \alpha_k$ let $\lambda_{s1} = [\lambda_s]$, otherwise let $\lambda_{s2} = [\lambda_s]$. Then John and Street (1992) show that, under the assumptions made about the range of concurrences in the component designs, minimizing (6.11) subject to (6.12) gives

$$n_{13} = n_{22} = n_{23} = 0, \quad n_{11} = n_{k1} - w, \quad n_{12} = w, \quad n_{21} = n_{k2}$$

if $w < n_{k1}$, and

$$n_{11} = n_{22} = n_{23} = 0, \quad n_{12} = 2n_{k1} - w, \quad n_{13} = w - n_{k1}, \quad n_{21} = n_{k2}$$

if $n_{k1} \le w \le 2n_{k1}$, and

$$n_{11} = n_{12} = n_{23} = 0, \quad n_{13} = n_{k1}, \quad n_{22} = w - 2n_{k1}, \quad n_{21} = n_{k2} - n_{22}$$

if $w > 2n_{k1}$, where

$$n_{k1} = v(v-1)(1 - \alpha_k), \qquad n_{k2} = v(v-1)\alpha_k$$

and

$$w = \begin{cases} v(v-1)\alpha_s & \text{if } \lambda_{s1} = [\lambda_s] \\ v(v-1)(1 + \alpha_s) & \text{if } \lambda_{s2} = [\lambda_s] \end{cases}$$

Using these values of n_{ij} leads to a lower bound for S_2 and hence to the upper bound U_1.

This lower bound for S_2 could also be substituted in (2.29) to give a further bound U_2.

The assumptions made by John and Street (1992) on the ranges of the treatment concurrences in D_k and D_s guarantee that the component designs will be good, but do not necessarily mean that U_1 is a global upper bound for a resolvable row–column design. There will be some designs for which the component designs must necessarily have more than two or three different concurrences. For example, the best resolvable block design for $v = 14$, $k = 7$, $s = 2$ and $r = 7$ has concurrences of 1, 2, 3, 4 and 5. John and Street

Table 6.3. *Value of x (y) for the upper bound* $U_1^* = U_1 + 0.001x$
$(U_2^* = U_2 + 0.001y)$

| | | | | r | | | |
s	4	5	6	7	8	9	10
2	18(18)	9(12)	16(19)	11(12)	11(12)	10 (9)	11(12)
3		7 (9)	8(10)	5 (8)	8(10)	7(10)	5 (8)
4			3 (4)	5 (8)	4 (7)	3 (5)	4 (6)
5				2 (2)	3 (4)	4 (5)	3 (5)
6					1 (1)	2 (3)	3 (4)
7						1 (1)	1 (2)
8							1 (1)

(1993) have used a mathematical optimization program to obtain
general solutions of (6.11) for $v \leq 500$, $2 \leq r \leq 10$, $2 \leq k \leq 10$ and
$k > s$. As a result U_1 has been shown to be a global upper bound
for the majority of parameter values. For $4 \leq s + 1 < r \leq 10$,
however, John and Street provide a simple adjustment factor to
U_1 to produce a general upper bound U_1^*. The bound is given by
$U_1^* = U_1 + 0.001x$ where x is given in Table 6.3; where there is no
entry in the table then $U_1^* = U_1$.

A similar adjustment can be made to U_2 to give $U_2^* = U_2 + 0.001y$
where the value of y is given in brackets in Table 6.3. For any
particular combination of parameters, the choice of upper bound
for the average efficiency factor is taken as the smaller of $\bar{e}$, U_0, U_1^*
and U_2^*.

Example 6.9
Suppose an upper bound on E is required for a resolvable row-
column design with $v = 20$, $k = 5$, $s = 4$ and $r = 4$. The arithmetic
mean bound is $\bar{e} = 0.6316$. From the results in Section 4.10,
$U_k = 0.7686$ and $U_s = 0.8187$ so that, from (5.16), $U_0 = 0.6568$.

Calculations for the John and Street (1992) bound give $n_{k1} =$
140, $n_{k2} = 240$ and $w = 320$ giving $n_{11} = n_{12} = n_{23} = 0$,
$n_{13} = 140$, $n_{21} = 200$ and $n_{22} = 40$. Using these values leads
to a lower bound $S_2 = 0.1586$ and to $U_1^* = U_1 = 0.6232$ and
$U_2^* = U_2 = 0.6230$. In this case, the tightest bound is given by U_2^*.

Example 6.10
For $v = 35$, $k = 7$, $s = 5$ and $r = 8$, $\bar{e} = 0.7059$, $U_0 = 0.7380$,

$U_1 = 0.7035$, $U_1^* = 0.7065$, $U_2 = 0.7036$ and $U_2^* = 0.7076$. Thus, $\bar{e}$ gives the tightest bound.

CHAPTER 7

Recovery of inter-block information

7.1 Introduction

The analysis of block designs presented in Chapter 1 was concerned with the estimation of treatment contrasts from comparisons made *within* blocks. If the blocks can be regarded as a random sample of blocks from some population then estimates of treatment contrasts may also be available from comparisons *between* blocks. Whether it is worth incorporating the information from this inter-block analysis with that from the within-block, or intra-block, analysis will depend on the extent to which the grouping of the experimental units into blocks has achieved a marked reduction in the error mean square. If this grouping is successful, i.e. block effects are large, then the amount of information which may be recovered from the inter-block analysis will be small. On the other hand if block effects are small there may be substantial gains to be achieved by recovering the inter-block information.

Similarly for row–column designs, there may be worthwhile gains to be achieved by recovering treatment information, where available, from comparisons made between rows and between columns. This chapter will be primarily concerned with the recovery of inter-block information from block and row–column designs. However, a general treatment of the subject will be outlined; also, in Section 7.8 a comparison with methods of analysis using neighbour models is given.

7.2 Orthogonal block structure

It is first necessary to consider the blocking structure of the experiment in order to determine what blocking comparison can be made, e.g. between blocks, within blocks, between columns, etc. The block structure derives from the use of plots which have an internal structure regardless of which treatments are applied to them. For simplicity attention will first be restricted to what Nelder (1965) calls *simple block structures*. For a block design this means that each block must contain the same number of plots, while in a $k \times s$ row–column design each row must contain s plots and each column k plots. In Section 7.7 the notation and discussion will be generalized to include unequal block sizes.

For a block design with b blocks each containing k plots, let y_{ij} represent the observation or yield on the ith plot of the jth block. Let $\bar{y}_j$ be the jth block mean and $\bar{y}$ the overall mean. Now consider the following *yield identity*:

$$y_{ij} = \bar{y} + (\bar{y}_j - \bar{y}) + (y_{ij} - \bar{y}_j) \qquad (7.1)$$

$$(i = 1, 2, \ldots, k; \; j = 1, 2, \ldots, b)$$

The last two terms in (7.1) represent contrasts between blocks and within blocks respectively.

A matrix representation of this identity can be given by letting the observations be set out in a vector $\mathbf{y}$ such that the first k observations are from the first block, the next k observations are from the second block, and so on. Then (7.1) can be written as

$$\mathbf{y} = \mathbf{C}_0\mathbf{y} + \mathbf{C}_1\mathbf{y} + \mathbf{C}_2\mathbf{y} \qquad (7.2)$$

where

$$
\begin{aligned}
\mathbf{C}_0 &= \mathbf{K}_b \otimes \mathbf{K}_k \\
\mathbf{C}_1 &= (\mathbf{I}_b - \mathbf{K}_b) \otimes \mathbf{K}_k \\
\mathbf{C}_2 &= \mathbf{I}_b \otimes (\mathbf{I}_k - \mathbf{K}_k)
\end{aligned}
$$

and where $\mathbf{K}_n$ is an $n \times n$ matrix with each element equal to $(1/n)$. Premultiplying a vector by $\mathbf{K}$ results in each element of the vector being replaced by its mean, whereas premultiplying by $\mathbf{I} - \mathbf{K}$ results in the mean being subtracted from each element in the vector. Hence, premultiplying $\mathbf{y}$ by $\mathbf{C}_0$ can be seen as an operation which first averages over plots and then over blocks to give a vector with each element equal to $\bar{y}$. Premultiplying $\mathbf{y}$ by $\mathbf{C}_1$ first averages over plots to give block means and then subtracts the overall mean from each block mean. Finally, premultiplying $\mathbf{y}$ by $\mathbf{C}_2$ results in block means being subtracted from the observations.

These $\mathbf{C}$ matrices form a *complete binary set* in that they are symmetric ($\mathbf{C}_i' = \mathbf{C}_i$), idempotent ($\mathbf{C}_i^2 = \mathbf{C}_i$), orthogonal ($\mathbf{C}_i\mathbf{C}_j = \mathbf{0}, \, i \neq j$) and satisfy $\mathbf{C}_0 + \mathbf{C}_1 + \mathbf{C}_2 = \mathbf{I}$.

A block structure which gives rise to the yield identity given in (7.1) or (7.2) is said to be *orthogonal*. The **C** matrices define the *strata* of this orthogonal block structure. The three strata are called respectively the *mean stratum*; the *block stratum*, representing between block-differences; and the *plots within blocks stratum*, representing between-plot, within-block differences.

The yield identity leads to a partitioning of the (uncorrected) total sum of squares into three components which correspond to the strata. From (7.1) the partition is

$$\sum\sum y_{ij}^2 = bk\bar{y}^2 + k\sum(\bar{y}_j - \bar{y})^2 + \sum\sum(y_{ij} - \bar{y}_j)^2 \quad (7.3)$$

or, in matrix notation using (7.2),

$$\mathbf{y}'\mathbf{y} = \mathbf{y}'\mathbf{C}_0\mathbf{y} + \mathbf{y}'\mathbf{C}_1\mathbf{y} + \mathbf{y}'\mathbf{C}_2\mathbf{y} \quad (7.4)$$

The degrees of freedom associated with the sum of squares $\mathbf{y}'\mathbf{C}_i\mathbf{y}$ are given by $\text{rank}(\mathbf{C}_i) = \text{trace}(\mathbf{C}_i)$, as $\mathbf{C}_i$ is idempotent; see (A.12). Thus, the total degrees of freedom $n = bk$ can be partitioned to give

$$bk = 1 + (b - 1) + b(k - 1) \quad (7.5)$$

These results are summarized in the null analysis of variance given in Table 7.1; it is called null since it is the analysis of variance that would be obtained if all plots received the same treatment. Usually the mean stratum is included in the total to give a (corrected) total sum of squares based on $bk - 1$ degrees of freedom.

The blocking structure of a row–column design can also be developed in a similar way. Let y_{ij} be the observation on the plot in the ith row and jth column, and let $\bar{y}_{i.}$ and $\bar{y}_{.j}$ represent ith row and jth column means respectively. The appropriate yield identity now becomes

Table 7.1. *Null analysis of variance for block designs*

Stratum	d.f.	s.s.
Mean	1	$\mathbf{y}'\mathbf{C}_0\mathbf{y}$
Block	$b - 1$	$\mathbf{y}'\mathbf{C}_1\mathbf{y}$
Plots within blocks	$b(k - 1)$	$\mathbf{y}'\mathbf{C}_2\mathbf{y}$
Total	bk	$\mathbf{y}'\mathbf{y}$

$$y_{ij} = \bar{y} + (\bar{y}_{i.} - \bar{y}) + (\bar{y}_{.j} - \bar{y}) + (y_{ij} - \bar{y}_{i.} - \bar{y}_{.j} + \bar{y})$$

$$(i = 1, 2, \ldots, k; \, j = 1, 2, \ldots, s)$$

In matrix notation the identity is

$$\mathbf{y} = \mathbf{C}_0 \mathbf{y} + \mathbf{C}_1 \mathbf{y} + \mathbf{C}_2 \mathbf{y} + \mathbf{C}_3 \mathbf{y} \tag{7.6}$$

where, if the observations are ordered row by row,

$$
\begin{aligned}
\mathbf{C}_0 &= \mathbf{K}_k \otimes \mathbf{K}_s \\
\mathbf{C}_1 &= (\mathbf{I}_k - \mathbf{K}_k) \otimes \mathbf{K}_s \\
\mathbf{C}_2 &= \mathbf{K}_k \otimes (\mathbf{I}_s - \mathbf{K}_s) \\
\mathbf{C}_3 &= (\mathbf{I}_k - \mathbf{K}_s) \otimes (\mathbf{I}_s - \mathbf{K}_s)
\end{aligned}
$$

It can be verified that these $\mathbf{C}$ matrices form a complete binary set.

The operations of premultiplying $\mathbf{y}$ by $\mathbf{C}_0$, $\mathbf{C}_1$, $\mathbf{C}_2$ and $\mathbf{C}_3$ are similar to those given above for block designs, where averaging and differencing are now carried out over rows and columns.

The four strata in a row–column design are respectively the mean stratum, the row stratum, the column stratum and the *row by column stratum*, involving interaction contrasts between rows and columns.

Sums of squares and degrees of freedom can be partitioned to give the null analysis of variance in Table 7.2.

A block design consists of two factors in which one factor (plots) is nested in the other (blocks). A row–column design has its two factors rows and columns cross-classified. Many other blocking structures will have the experimental plots grouped into blocks in ways which can be described in terms of nesting and crossing. For instance, a resolvable row–column design, such as a lattice square

Table 7.2. *Null analysis of variance for row–column designs*

Stratum	d.f.	s.s.
Mean	1	$\mathbf{y}'\mathbf{C}_0\mathbf{y}$
Row	$k - 1$	$\mathbf{y}'\mathbf{C}_1\mathbf{y}$
Column	$s - 1$	$\mathbf{y}'\mathbf{C}_2\mathbf{y}$
Row by column	$(k - 1)(s - 1)$	$\mathbf{y}'\mathbf{C}_3\mathbf{y}$
Total	ks	$\mathbf{y}'\mathbf{y}$

design, has rows crossed with columns which are then nested within replicates. For a simple orthogonal blocking structure the essential requirement is that the blocks all contain the same number of plots. Nelder (1965) gives simple rules for obtaining the yield identity and null analysis of variance for any such structure.

An advantage of a simple block structure is that it has a complete randomization theory. The block structure is preserved under any permutation of the labelling of the plots. For instance, in a block design, the blocks may be labelled arbitrarily and the plots within blocks may also be labelled arbitrarily. There is no connection between the ith plot in the jth block and the ith plot in the j'th block. This arbitrary labelling can be achieved only if all blocks contain the same number of plots. If the suffixes in $y_{ijk...}$ identify the levels of the blocking factors then the randomization procedure for any simple block structure consists of re-ordering the values of each suffix at random.

The population of all possible vectors of observations generated by the randomization procedure gives the null randomization distribution; again assuming the null experiment in which all plots receive the same treatment. The variance–covariance matrix of this distribution, the null analysis of variance and expectations of sums of squares can be derived from the form of the simple block structure. Details of this randomization theory can be found in Nelder (1965).

In the discussion of Bailey (1991), however, Preece points out that a disadvantage of the randomization analysis approach is that it does not accommodate the important practical situation of unequal block sizes. For example, Patterson and Hunter (1983) comment that in about half of the United Kingdom statutory cereal trials, block sizes of k_1 and k_2 plots are used, where $k_2 = k_1 - 1$. Some reasons for the use of unequal block sizes are given in Section 4.9. Another consideration is that in a field situation the incomplete blocks are often just a convenient grouping of plots and so there is considerable scope for choosing the size and number of blocks. Hence, for the more general incomplete block situation it is necessary to approach the analysis by declaring a statistical model. In the next section this model-based approach is discussed for equal block sizes and then generalized in Section 7.7.

7.3 Error covariance structure

Following from (1.3), and ignoring treatment parameters as treatments play no part in the blocking structure, an appropriate model for a null block design is

$$y_{ij} = \mu + \beta_j + \varepsilon_{ij} \qquad (i = 1, 2, \ldots, k; \; j = 1, 2, \ldots, b) \qquad (7.7)$$

It will be assumed that the errors ε_{ij} are uncorrelated random variables with zero means and $\operatorname{var}(\varepsilon_{ij}) = \sigma^2$, that the block effects β_j are also uncorrelated random variables with zero means and $\operatorname{var}(\beta_j) = \sigma_1^2$ and that the β_j are uncorrelated with the ε_{ij}, i.e. $\operatorname{cov}(\beta_j, \varepsilon_{ij}) = 0$. It then follows that

$$\operatorname{var}(y_{ij}) = \sigma^2 + \sigma_1^2$$

$$\operatorname{cov}(y_{ij}, y_{i'j'}) = \begin{cases} \sigma_1^2, & j = j', i \neq i' \\ 0, & j \neq j' \end{cases}$$

Thus two observations from the same block have covariance σ_1^2 whereas two observations from different blocks have zero covariance. The quantities σ^2 and σ_1^2 are called *variance components* for plots and blocks respectively.

In matrix notation, the variance–covariance matrix of the y_{ij} is a block diagonal matrix which can be written as

$$V(\mathbf{y}) = \mathbf{V} = \mathbf{I}_b \otimes \{(\sigma^2 + \sigma_1^2)\mathbf{I}_k + \sigma_1^2(\mathbf{J}_k - \mathbf{I}_k)\}$$

i.e.

$$\mathbf{V} = \mathbf{I}_b \otimes (\sigma^2 \mathbf{I}_k + k\sigma_1^2 \mathbf{K}_k) \qquad (7.8)$$

Now using the yield identity given in (7.2) it follows that

$$\mathbf{V} = (\mathbf{C}_0 + \mathbf{C}_1 + \mathbf{C}_2)' \mathbf{V} (\mathbf{C}_0 + \mathbf{C}_1 + \mathbf{C}_2)$$

and since it is easily shown that $\mathbf{C}_i \mathbf{V} = \xi_i \mathbf{C}_i \; (i = 0, 1, 2)$ where

$$\xi_0 = \sigma^2 + k\sigma_1^2, \quad \xi_1 = \sigma^2 + k\sigma_1^2, \quad \xi_2 = \sigma^2 \qquad (7.9)$$

then $\mathbf{V}$ can be written as

$$\mathbf{V} = \xi_0 \mathbf{C}_0 + \xi_1 \mathbf{C}_1 + \xi_2 \mathbf{C}_2 \qquad (7.10)$$

Note that the columns of $\mathbf{C}_i$ are eigenvectors of $\mathbf{V}$ with eigenvalues ξ_i. Hence (7.10) represents a canonical or spectral decomposition of $\mathbf{V}$, which is then said to be in *spectral form*. Since the $\mathbf{C}_i$ matrices defined the strata of the blocking structure, the ξ_i are called the *stratum variances*.

A matrix representation of (7.7) is, following (1.5), given by

$$\mathbf{y} = \mathbf{1}\mu + \mathbf{Z}\beta + \varepsilon$$

where $\mathbf{Z}$ is the design matrix for blocks. With $V(\beta) = \sigma_1^2\mathbf{I}$, $V(\varepsilon) = \sigma^2\mathbf{I}$ and $\text{cov}(\beta, \varepsilon) = \mathbf{0}$,

$$\mathbf{V} = \sigma^2\mathbf{I} + \sigma_1^2\mathbf{ZZ}' \tag{7.11}$$

which can be shown to be the same as (7.10) by noting that

$$\mathbf{ZZ}' = k(\mathbf{I}_b \otimes \mathbf{K}_k) \tag{7.12}$$

For a row–column design, the null model is given by

$$y_{ij} = \mu + \rho_i + \gamma_j + \varepsilon_{ij} \tag{7.13}$$

where ρ_i and γ_j are row and column effects respectively. It is now assumed that the ρ_i, γ_j and ε_{ij} are random variables with zero means and with $\text{var}(\varepsilon_{ij}) = \sigma^2$, $\text{var}(\rho_i) = \sigma_1^2$ and $\text{var}(\gamma_j) = \sigma_2^2$. Further all random variables are uncorrelated with each other. Then

$$\text{var}(y_{ij}) = \sigma^2 + \sigma_1^2 + \sigma_2^2$$

$$\text{cov}(y_{ij}, y_{i'j'}) = \begin{cases} \sigma_1^2, & i = i', j \neq j' \\ \sigma_2^2, & i \neq i', j = j' \\ 0, & i \neq i', j \neq j' \end{cases}$$

Thus the covariance between any two observations is σ_1^2, σ_2^2 or 0 depending on whether they are in the same row, same column or different rows and columns respectively. The quantities σ^2, σ_1^2 and σ_2^2 are the variance components for plots, rows and columns respectively.

In matrix notation, the variance–covariance matrix of y_{ij} is given by

$$\mathbf{V} = \mathbf{I}_k \otimes \{(\sigma^2 + \sigma_1^2 + \sigma_2^2)\mathbf{I}_s + \sigma_1^2(\mathbf{J}_s - \mathbf{I}_s)\} + (\mathbf{J}_k - \mathbf{I}_k) \otimes \sigma_2^2\mathbf{I}_s$$

i.e.

$$\mathbf{V} = \mathbf{I}_k \otimes (\sigma^2\mathbf{I}_s + s\sigma_1^2\mathbf{K}_s) + k\sigma_2^2\mathbf{K}_k \otimes \mathbf{I}_s$$

Using the $\mathbf{C}_i$ matrices given in (7.6) it can then be established that the spectral form of $\mathbf{V}$ is given by

$$\mathbf{V} = \xi_0\mathbf{C}_0 + \xi_1\mathbf{C}_1 + \xi_2\mathbf{C}_2 + \xi_3\mathbf{C}_3 \tag{7.14}$$

where the stratum variances are

$$\begin{aligned} \xi_0 &= \sigma^2 + s\sigma_1^2 + k\sigma_2^2 \\ \xi_1 &= \sigma^2 + s\sigma_1^2 \\ \xi_2 &= \sigma^2 + k\sigma_2^2 \\ \xi_3 &= \sigma^2 \end{aligned} \tag{7.15}$$

In summary, for any simple orthogonal block structure the yield identity can be represented by

$$y = \sum C_i y \qquad (7.16)$$

where the C_i matrices form a complete binary set and define the strata of the blocking structure. The null analysis of variance is given by the *quadratic identity*

$$y'y = \sum y'C_i y \qquad (7.17)$$

and the *error covariance structure* by

$$V = \sum \xi_i C_i \qquad (7.18)$$

where the ξ_i are the stratum variances.

7.4 Generalized least squares analysis

Sections 7.2 and 7.3 were concerned with the null experiment in which all plots receive the same treatment. In the null model, therefore, each observation has the same expected value, namely μ. When different treatments, either as a single set of treatments or as combinations of a number of treatment factors, are applied to the plots then the expected value of the observation from a particular plot will depend on which treatment is applied to that plot. Since the error covariance structure depends only on the blocking structure, the model is thus given by

$$\begin{aligned} E(y) &= X\tau \\ V(y) &= V \end{aligned} \qquad (7.19)$$

where X is the design matrix for treatments, τ is the vector of treatment parameters and where V is given by (7.18).

Generalized least squares analysis can be used to provide an estimator $\hat{\tau}$ of τ. The estimator is obtained as a solution to the generalized least squares equations

$$X'V^{-1}X\hat{\tau} = X'V^{-1}y$$

Since V is in spectral form it follows that

$$V^{-1} = \sum \xi_i^{-1} C_i$$

Hence the generalized least squares equations become

$$\sum \xi_i^{-1} X'C_i X\hat{\tau} = \sum \xi_i^{-1} XC_i y \qquad (7.20)$$

The variance–covariance matrix of the estimator is then given by

$$V(\hat{\tau}) = \left(\sum \xi_i^{-1} \mathbf{X}' \mathbf{C}_i \mathbf{X}\right)^- \qquad (7.21)$$

The matrix $\mathbf{A}_i = \mathbf{X}' \mathbf{C}_i \mathbf{X}$ is called the treatment *information matrix* in the ith stratum.

7.5 Estimation of stratum variances

If the stratum variances ξ_i are not known, as will usually be the case in practice, then estimates of them will be required in order to obtain $\hat{\tau}$ and $V(\hat{\tau})$ from (7.20) and (7.21) respectively. A number of methods of estimation have been given and some of them will be briefly discussed in this section.

In the intra-block analysis of variance given in Table 1.3 it was stated that the residual mean square s^2 provided an unbiased estimator of the variance σ^2. That is, the estimator of σ^2 is obtained by equating the residual mean square in the plots within blocks stratum to its expectation. More generally, an estimator of the ith stratum variance ξ_i can be obtained by equating the residual mean square in the ith stratum to its expectation. The expected value of this mean square will now be obtained.

The treatment information in the ith stratum will be obtained from comparisons made among the elements of the vector $\mathbf{z}_i = \mathbf{C}_i \mathbf{y}$. From (7.19), it follows that $E(\mathbf{z}_i) = \mathbf{C}_i \mathbf{X} \tau$ and $V(\mathbf{z}_i) = \xi_i \mathbf{C}_i$, so that the least squares estimator $\hat{\tau}_i$ of τ in the ith stratum is

$$\hat{\tau}_i = \mathbf{A}_i^- \mathbf{X}' \mathbf{C}_i \mathbf{y} \qquad (7.22)$$

Hence the residual mean square in the ith stratum is given by

$$\begin{aligned}
\text{RSS}_i &= (\mathbf{z}_i - \mathbf{C}_i \mathbf{X} \hat{\tau}_i)'(\mathbf{z}_i - \mathbf{C}_i \mathbf{X} \hat{\tau}_i) \\
&= \mathbf{z}_i'(\mathbf{I} - \mathbf{X} \mathbf{A}_i^- \mathbf{X}') \mathbf{z}_i \qquad (7.23)
\end{aligned}$$

The expected value of RSS_i is

$$E(\text{RSS}_i) = \text{trace}\{E(\text{RSS}_i)\} = \text{trace}\{(\mathbf{I} - \mathbf{X} \mathbf{A}_i^- \mathbf{X}') E(\mathbf{z}_i \mathbf{z}_i')\}$$

By definition $V(\mathbf{z}_i) = E\{(\mathbf{z}_i - \mathbf{C}_i \mathbf{X} \tau)(\mathbf{z}_i - \mathbf{C}_i \mathbf{X} \tau)'\}$ so that

$$E(\mathbf{z}_i \mathbf{z}_i') = \xi_i \mathbf{C}_i + \mathbf{C}_i \mathbf{X} \tau (\mathbf{C}_i \mathbf{X} \tau)'$$

Now

$$\begin{aligned}
\text{trace}\{(\mathbf{I} - \mathbf{X} \mathbf{A}_i^- \mathbf{X}') \mathbf{C}_i \mathbf{X} \tau (\mathbf{C}_i \mathbf{X} \tau)'\} = \\
\text{trace}\{(\mathbf{C}_i \mathbf{X} \tau)'(\mathbf{I} - \mathbf{X} \mathbf{A}_i^- \mathbf{X}') \mathbf{C}_i \mathbf{X} \tau\} = 0
\end{aligned}$$

since $\mathbf{X}'\mathbf{C}_i\mathbf{X}\mathbf{A}_i^-\mathbf{X}'\mathbf{C}_i\mathbf{X} = \mathbf{X}'\mathbf{C}_i\mathbf{X}$. Hence, using (A.20),

$$
\begin{aligned}
\mathrm{E}(\mathrm{RSS}_i) &= \xi_i\{\mathrm{trace}(\mathbf{C}_i) - \mathrm{trace}(\mathbf{X}'\mathbf{C}_i\mathbf{X}\mathbf{A}_i^-)\} \\
&= \xi_i\{\mathrm{rank}(\mathbf{C}_i) - \mathrm{rank}(\mathbf{A}_i)\} \\
&= \xi_i d_i
\end{aligned}
\tag{7.24}
$$

where d_i is therefore the degrees of freedom associated with RSS_i. Hence, the residual mean square RSS_i/d_i from the ith stratum gives an unbiased estimator of the stratum variance ξ_i.

The disadvantage of this method is that the residual mean square in the ith stratum may be based on an inadequate number of degrees of freedom, i.e. d_i can often be small, zero or even negative, as is the case in block designs with $v > b$. The reason for this is that although there may only be partial information on certain contrasts in the ith stratum each independent contrast will account for a whole degree of freedom in calculating d_i. To overcome this difficulty Nelder (1968) advocates equating the *actual* residual mean square in the ith stratum to its expectation, i.e. using $\hat{\boldsymbol{\tau}}$ from (7.20) in (7.23) in place of the ith stratum estimator $\hat{\boldsymbol{\tau}}_i$. Both the residual mean square and its expectation will then involve the stratum variances so that an iterative procedure will be required to calculate the estimate of ξ_i. Nelder (1968) gives details of how this can be carried out for the class of *generally balanced* designs, i.e. designs for which the information matrices $\mathbf{A}_i = \mathbf{X}'\mathbf{C}_i\mathbf{X}$ are spanned by a common set of eigenvectors.

Alternatively the statistical likelihood of the actual residuals can be maximized to obtain estimates $\hat{\xi}_i$ of the stratum variances. The $\hat{\xi}_i$ can then be substituted into $\mathbf{V}$ to provide a generalized least squares solution for $\boldsymbol{\tau}$. This method is known as residual maximum likelihood (REML). Patterson and Thompson (1971) give a method for REML estimation of stratum variances in block designs, which gives identical results to those of Nelder (1968) when block sizes are equal. The REML method of estimation is computationally intensive, often requiring several iterations for the estimates to converge. Fortunately the procedure is included in the latest versions of the statistical software packages GENSTAT and SAS. One big advantage of REML estimation in these packages is that a general method is provided for estimating variance components and treatment parameters for a wide range of blocking structures such as row–column designs and Latinized designs.

In recovering inter-block information from balanced incomplete block and lattice designs, Yates (1939, 1940b) estimated the stratum variances by equating the adjusted block mean square and residual mean square to their expectations and showed that such a procedure resulted in little loss of efficiency. In the next section this method is demonstrated for block designs. The method can be extended to block designs in general and to row–column designs but each different blocking structure requires a separate algebraic development of the stratum variance estimators. On the other hand when suitable computer software is available, REML provides a convenient, efficient and flexible method of estimation.

It should be noted that if there is no treatment information available in the ith stratum then the combined estimate obtained from (7.20) will be equal to that obtained by setting $\xi_i^{-1} = 0$. For example, in a complete block design all the treatment information is available in the plots within blocks stratum. Hence, the intra-block estimate will be the same as the combined estimate $\hat{\tau}$ no matter what estimate of ξ_1 is used. Finally, if it is unreasonable to consider the blocking effects in any stratum to be random variables then the treatment information in the stratum concerned is ignored, by again setting the reciprocal of the stratum variance to zero.

7.6 Block designs

7.6.1 Combined treatment estimates

Let v treatments be set out in a block design consisting of b blocks each of k plots such that each treatment is replicated r times. The intra-block analysis has been considered in Chapter 1. To recover inter-block information it will be assumed that the block effects β_j and the error terms ε_{ij} are random variables with $E(\beta_j) = 0$, $\text{var}(\beta_j) = \sigma_1^2$ and $E(\varepsilon_{ij}) = 0$, $\text{var}(\varepsilon_{ij}) = \sigma^2$ and that all random variables are uncorrelated. The appropriate model for the combined analysis is then given by

$$E(\mathbf{y}) = \mathbf{X}\tau, \quad V(\mathbf{y}) = \xi_0 \mathbf{C}_0 + \xi_1 \mathbf{C}_1 + \xi_2 \mathbf{C}_2 \qquad (7.25)$$

where $\mathbf{X}$ is the $n \times v$ design matrix for treatments, where the $\mathbf{C}_i$ matrices are defined in (7.2) and where the stratum variances are

$$\xi_0 = \xi_1 = \sigma^2 + k\sigma_1^2, \quad \xi_2 = \sigma^2 \qquad (7.26)$$

The equations for the generalized least squares estimator of τ are then given by (7.20). The treatment information matrices are

$$
\begin{aligned}
\mathbf{A}_0 &= \mathbf{X}'\mathbf{C}_0\mathbf{X} = (r^2/n)\mathbf{J} \\
\mathbf{A}_1 &= \mathbf{X}'\mathbf{C}_1\mathbf{X} = (1/k)\mathbf{N}\mathbf{N}' - (r^2/n)\mathbf{J} \\
\mathbf{A}_2 &= \mathbf{X}'\mathbf{C}_2\mathbf{X} = r\mathbf{I} - (1/k)\mathbf{N}\mathbf{N}'
\end{aligned}
\tag{7.27}
$$

where use has been made of (1.6), (1.7) and the fact that, from (1.8) and (7.12),

$$
\mathbf{N}\mathbf{N}' = \mathbf{X}'\mathbf{Z}\mathbf{Z}'\mathbf{X} = k\mathbf{X}'(\mathbf{I}_b \otimes \mathbf{K}_k)\mathbf{X}
$$

Note that $\mathbf{A}_2$ is the information matrix $\mathbf{A}$ of the intra-block analysis given in (1.44). It is therefore the information matrix in the plots within blocks stratum.

Using (7.26) and (7.27)

$$
\begin{aligned}
\sum \xi_i^{-1}\mathbf{X}'\mathbf{C}_i\mathbf{X} &= (\xi_0^{-1}/k)\mathbf{N}\mathbf{N}' + \xi_2^{-1}\{r\mathbf{I} - (1/k)\mathbf{N}\mathbf{N}'\} \\
&= \xi_2^{-1}\left\{r\mathbf{I} - (1/k)\left(1 - \frac{\xi_2}{\xi_0}\right)\mathbf{N}\mathbf{N}'\right\}
\end{aligned}
$$

Now

$$
1 - \frac{\xi_2}{\xi_0} = 1 - \frac{\sigma^2}{\sigma^2 + k\sigma_1^2} = \frac{k\sigma_1^2}{\sigma^2 + k\sigma_1^2} = \frac{1}{1 + \eta^{-1}}
$$

where

$$
\eta = k\sigma_1^2/\sigma^2
\tag{7.28}
$$

Hence

$$
\sum \xi_i^{-1}\mathbf{X}'\mathbf{C}_i\mathbf{X} = \xi_2^{-1}\mathbf{A}_c
$$

where

$$
\mathbf{A}_c = r\mathbf{I} - (1/k^*)\mathbf{N}\mathbf{N}'
\tag{7.29}
$$

and

$$
k^* = k(1 + \eta^{-1})
\tag{7.30}
$$

If $\mathbf{T}$ and $\mathbf{B}$ are the vectors of treatment and block totals respectively then, using (1.8) and (1.9), it can be seen that

$$
\begin{aligned}
\mathbf{q}_0 &= \mathbf{X}'\mathbf{C}_0\mathbf{y} = r\bar{y}\mathbf{1} \\
\mathbf{q}_1 &= \mathbf{X}'\mathbf{C}_1\mathbf{y} = (1/k)\mathbf{N}\mathbf{B} - r\bar{y}\mathbf{1} \\
\mathbf{q}_2 &= \mathbf{X}'\mathbf{C}_2\mathbf{y} = \mathbf{T} - (1/k)\mathbf{N}\mathbf{B}
\end{aligned}
\tag{7.31}
$$

Note again that $\mathbf{q}_2$ is the vector of adjusted treatment totals of the intra-block analysis given in (1.15). It follows that

$$
\sum \xi_i^{-1}\mathbf{X}'\mathbf{C}_i\mathbf{y} = (\xi_0^{-1}/k)\mathbf{N}\mathbf{B} + \xi_2^{-1}\{\mathbf{T} - (1/k)\mathbf{N}\mathbf{B}\}
$$

i.e.

$$\sum \xi_i^{-1} X' C_i y = \xi_2^{-1} q_c$$

where

$$q_c = T - (1/k^*) NB \tag{7.32}$$

and k^* is given by (7.30). Hence, the generalized least squares equations are

$$A_c \hat{\tau} = q_c \tag{7.33}$$

where A_c and q_c are given by (7.29) and (7.32) respectively.

If $\eta^{-1} = 0$ then $k^* = k$ and the equations are those of the intra-block analysis given by (1.13). Thus, when the block variance component σ_1^2 is very large compared with the error variance σ^2 there will be little to be gained from recovering inter-block information. On the other hand if $\sigma_1^2 = 0$, then (7.33) reduces to $r\hat{\tau} = T$, i.e. the equations for a complete block analysis.

Recall that in Section 1.4 the columns of the intra-block treatment information matrix A were spanned by a set of normalized eigenvectors $p_1, p_2, \ldots, p_v$ with corresponding eigenvalues $\theta_1, \theta_2, \ldots, \theta_v$. Since this means that

$$NN' p_i = k(r - \theta_i) p_i$$

it follows that $A_c p_i = \theta_i^* p_i$ where

$$\theta_i^* = \{r(k^* - k) + k\theta_i\}/k^* \tag{7.34}$$

Hence, unless $k = k^*$, all the eigenvalues of A_c are non-zero so that A_c is a non-singular matrix. The combined treatment estimator $\hat{\tau}$ is then given by

$$\hat{\tau} = A_c^{-1} q_c \tag{7.35}$$

with variance–covariance matrix

$$V(\hat{\tau}) = A_c^{-1} \sigma^2 \tag{7.36}$$

Note that $\hat{\tau} 1 = v\bar{y}$ so that the mean of the combined treatment estimates is the overall mean $\bar{y}$, which means that they are directly comparable with the unadjusted treatment means and the intra-block estimates given by (1.20).

Estimates of σ^2 and σ_1^2 will be required to obtain $\hat{\tau}$. The residual mean square in the plots within blocks stratum gives an unbiased estimator of σ^2. The estimator of σ_1^2 will be obtained by equating the adjusted block sum of squares in the intra-block analysis to its expectation. This is the method of estimation used by Yates (1939, 1940b) and Cochran and Cox (1957) for balanced incomplete block and lattice designs.

From (1.37), the adjusted block sum of squares is given by

$$S(\beta/\mu, \tau) = \mathbf{s'L^- s} \qquad (7.37)$$

where

$$\mathbf{s} = \mathbf{B} - (1/r)\mathbf{N'T}$$

and $\mathbf{L}^-$ is a generalized inverse matrix of

$$\mathbf{L} = k\mathbf{I} - (1/r)\mathbf{N'N}$$

Now $\mathbf{s} = \mathbf{Z'Uy}$ and $\mathbf{L} = \mathbf{Z'UZ}$, where $\mathbf{U} = \mathbf{I} - (1/r)\mathbf{XX'}$ is an idempotent matrix, since $\mathbf{X'X} = r\mathbf{I}$. Then

$$\mathbf{s'L^- s} = \mathbf{y'UZ(Z'UZ)^- Z'Uy}$$

so that, using (7.11),

$$E(\mathbf{s'L^- s}) = \text{trace}\{\mathbf{UZ(Z'UZ)^- Z'U}(\sigma^2\mathbf{I} + \sigma_1^2\mathbf{ZZ'})\}$$

since $E(\mathbf{yy'}) = V(\mathbf{y}) + \mathbf{X}\tau(\mathbf{X}\tau)'$ and $\mathbf{X'U} = \mathbf{0}$. Now

$$\text{trace}\{\mathbf{UZ(Z'UZ)^- Z'U}\} =$$
$$\text{trace}\{\mathbf{(Z'UZ)^- Z'UZ}\} = \text{rank}(\mathbf{L}) = b - 1$$

and

$$\text{trace}(\mathbf{UZL^- Z'UZZ'}\} = \text{trace}(\mathbf{LL^- L}) = \text{trace}(\mathbf{L}) = v(r-1)$$

Hence

$$E(\mathbf{s'L^- s}) = (b-1)\sigma^2 + r(v-1)\sigma_1^2$$

Let RMS and BMS represent the residual mean square and adjusted block mean square respectively, then it follows that estimators of σ^2 and σ_1^2 are given by

$$\hat{\sigma}^2 = \text{RMS}, \qquad \hat{\sigma}_1^2 = \frac{(b-1)(\text{BMS} - \text{RMS})}{r(v-1)} \qquad (7.38)$$

Slightly different estimates are obtained for resolvable designs. Replicates are usually regarded as fixed while blocks within replicates are random variables. Hence, the adjusted block sum of squares, BSS, for resolvable designs will be

$$\text{BSS} = \mathbf{s'L^- s} - (\mathbf{R'R}/v - G^2/n) \qquad (7.39)$$

where $\mathbf{R}$ is the vector of replicate totals and G the overall total. In a similar way it can be shown that

$$E(\text{BSS}) = (b-r)\sigma^2 + (v-k)(r-1)\sigma_1^2$$

so that

$$\hat{\sigma}^2 = \text{RMS}, \qquad \hat{\sigma}_1^2 = \frac{r(\text{BMS} - \text{RMS})}{k(r-1)} \qquad (7.40)$$

Standard errors of differences between adjusted treatment means are obtained from the variance–covariance matrix given in (7.36), where variance components σ^2 and σ_1^2 are replaced by their estimates. Following (2.2), an approximate average standard error is given by

$$\text{ASE} = \sqrt{(2\hat{\sigma}^2/rE^*)} \qquad (7.41)$$

where

$$E^* = \{(k^* - k) + kE\}/k^*$$

is given by replacing each θ_i in (7.34) by rE, where E is the average efficiency factor.

7.6.2 Example of the analysis of a block design

The results of a spring oats trial grown in Craibstone, near Aberdeen, have been kindly made available by the North of Scotland College of Agriculture. The trial involved 24 varieties and three replicates, each consisting of six blocks of four plots. Thus the design parameters are $v = 24$, $r = 3$, $b = 18$ and $k = 4$. The resolvable block design used was an $\alpha(0, 1)$-design obtained from the α-array

$$\begin{array}{ccc} 0 & 0 & 0 \\ 0 & 1 & 5 \\ 0 & 3 & 2 \\ 0 & 2 & 3 \end{array}$$

as discussed in Section 4.4. The α-design has an average efficiency factor of $E = 0.7265$.

The varieties, which were numbered $1, 2, \ldots, 24$, were randomly assigned to the treatment labels $0, 1, \ldots, 23$ in the α-design. Replicates, blocks within replicates and plots within blocks were also randomly permuted. The design after randomization is as follows:

Table 7.3. *Dry matter grain yield (tonne/ha)*

Replicate 1					
4.1172	4.6540	4.2323	4.2530	4.7876	4.7085
4.4461	4.1736	4.7572	3.3420	5.0902	5.2560
5.8757	4.0141	4.4906	4.7269	4.1505	4.9577
4.5784	4.3350	3.9737	4.9989	5.1202	3.3986

Replicate 2					
3.9926	3.9039	5.3127	5.1202	5.1566	5.3148
3.6056	4.9114	5.1163	4.2955	5.0988	4.6297
4.5294	3.7999	5.3802	4.9057	5.4840	5.1751
4.3599	4.3042	5.0744	5.7161	5.0969	5.3024

Replicate 3					
3.9205	4.0510	4.3234	4.1746	4.4130	2.8873
4.6512	4.6783	4.2486	4.7512	4.2397	4.1972
4.3887	3.1407	4.3960	4.0875	4.3852	3.7349
4.5552	3.9821	4.2474	3.8721	3.5655	3.6096

Replicate		
1	**2**	**3**
11 21 23 13 17 6	8 24 12 5 2 19	11 2 17 12 21 3
4 10 14 3 15 12	20 15 11 9 18 7	1 15 18 13 22 5
5 20 16 19 7 24	14 3 21 10 13 6	14 9 4 10 16 20
22 2 18 8 1 9	4 23 17 1 22 16	19 8 6 23 24 7

The plots were, in fact, laid out in a single line of 72, with variety 11 on the 1st plot, 4 on the 2nd, 5 on the 3rd,..., and 7 on the 72nd. The dry matter grain yields (in tonne/ha) are given in Table 7.3, where each yield can be identified with a variety according to the above design. For instance, the yield of variety 14 on plot 2 of block 3 in replicate 1 is 4.7572.

The intra-block analysis of variance is given in Table 7.4. A comparison of the adjusted varieties mean square with the residual mean square gives an F value of 5.24 on 23 and 31 degrees of freedom, which is highly significant. Thus there are significant differences between varieties. The adjusted intra-block variety means, together with the average standard error of differences between adjusted means (SED), are given in Table 7.5. The plot

Table 7.4. *Analysis of variance table*

Source of variation	d.f.	s.s.	m.s.
Replicates	2	6.1355	3.0677
Blocks within replicates (unadj.)	15	7.6182	0.5079
Varieties (adj.)	23	10.0619	0.4375
Blocks within replicates (adj.)	15	3.6036	0.2402
Varieties (unadj.)	23	14.0765	0.6120
Residual	31	2.5874	0.0835
Total	71	26.4030	

of residuals against fitted values did not indicate any obvious departures from the assumptions underlying the analysis.

The estimates of σ^2 and σ_1^2 are $\hat{\sigma}^2 = 0.0835$ and $\hat{\sigma}_1^2 = 0.0588$ respectively. The adjusted combined variety means are also given in Table 7.5; the combined estimates have been arranged in order of magnitude to facilitate the comparison of varieties. From (7.41) the approximate average standard error of differences is 0.264, corresponding to a value of $E^* = 0.798$. The exact average standard error of differences is 0.265 with a range from 0.258 to 0.270. In this example the results using REML estimation are very similar. The attraction of REML, however, is that the computation can be readily carried out in GENSTAT or SAS.

The gain in *efficiency* resulting from dividing the 24 plots within each replicate into six blocks of four can be calculated as follows. If blocks within replicates are ignored, the residual mean square from the resulting complete block analysis would be given by pooling the adjusted block mean square of 0.2402 and the residual mean square of 0.0835. This gives a complete block residual mean square of

$$\text{CMS} = \frac{15 \times 0.2402 + 31 \times 0.0835}{46} = 0.1346$$

so that the standard error of mean differences would have been $\sqrt{(2 \times 0.1346/3)} = 0.300$; compared with 0.265 for the combined analysis of the α-design. Note that using a computer package, CMS can easily be obtained by constructing the complete block analysis

Table 7.5. *Adjusted variety means*

Variety	Complete block estimates	Intra-block estimates	Combined estimates
1	5.16	5.08	5.11
5	5.06	5.03	5.04
15	4.89	5.02	4.97
19	4.87	4.84	4.84
21	4.82	4.76	4.80
14	4.56	4.90	4.77
13	4.83	4.73	4.76
12	4.91	4.64	4.76
16	4.73	4.72	4.73
17	4.73	4.51	4.60
6	4.71	4.43	4.54
22	4.64	4.46	4.53
8	4.32	4.67	4.53
4	4.40	4.54	4.49
2	4.51	4.47	4.48
10	4.39	4.36	4.37
18	4.44	4.32	4.36
11	4.38	4.22	4.28
23	4.14	4.31	4.25
24	4.14	4.14	4.15
7	4.13	4.11	4.11
20	3.78	4.20	4.04
9	3.61	3.44	3.50
3	3.34	3.61	3.50
SED	0.300	0.277	0.265

of variance table as in Table 1.1. The efficiency is then

$$\text{Efficiency} = \left(\frac{0.300}{0.265}\right)^2 = 1.282$$

Thus there has been a 28.2% increase in efficiency resulting from the use of blocks within replicates.

The Craibstone trial was one of many similar trials covering different parts of Scotland and a range of seasonal conditions. The results of this analysis (adjusted variety means and standard error) went forward to be co-ordinated with the results of other trials.

7.7 Resolvable designs with unequal block sizes

In practice incomplete block designs with unequal block sizes may have to be used. This is particularly the case for resolvable designs when the number of treatments does not factorize in the form $v = ks$. Patterson and Williams (1976a) suggested the use of two block sizes k_1 and k_2, with $k_2 = k_1 - 1$, and proposed a method of construction based on α-designs; see Section 4.9. Such designs can be easily generated using the software package ALPHA+.

The randomization analysis approach of Nelder (1965) does not cover the case of unequal block sizes. Instead the models used in previous sections will need to be modified. To cater for replicate effects and unequal block sizes, the model (7.25) can be generalized to

$$E(\mathbf{y}) = \mathbf{R}\boldsymbol{\pi} + \mathbf{X}\boldsymbol{\tau}, \quad V(\mathbf{y}) = \mathbf{V} = \sigma^2(\mathbf{I} + \eta\mathbf{P}_B) \tag{7.42}$$

where η is a parameter, $\boldsymbol{\pi}$ is an $r \times 1$ vector of replicate parameters, $\mathbf{R}$ is the $n \times r$ design matrix for replicates, $\mathbf{P}_B$ is the *projection matrix* for blocks and is defined as $\mathbf{P}_B = \mathbf{Z}(\mathbf{Z}'\mathbf{Z})^{-1}\mathbf{Z}'$ where $\mathbf{Z}$ is the design matrix for blocks. Projection matrices are idempotent matrices and form convenient general algebraic tools for describing the influence of blocking and treatment factors.

From (1.7), $\mathbf{Z}'\mathbf{Z} = \mathbf{k}^\delta$ is a diagonal matrix of block sizes, with jth block size k_j ($j = 1, 2, \ldots, b$). Hence if the block sizes are equal the variance matrix $\mathbf{V}$ in (7.42) reduces to that in (7.11) with η given by (7.28).

Note that if the observations in $\mathbf{y}$ are ordered block by block as in (7.2), the projection matrix $\mathbf{P}_B$ is a block diagonal matrix with jth diagonal submatrix $\mathbf{K}_{k_j}$; a generalization of (7.12). It then follows from model (7.42) that

$$
\begin{aligned}
\mathrm{var}(y_{ij}) &= \sigma^2(1 + \eta/k_j) \\
\mathrm{cov}(y_{ij}, y_{i'j'}) &= \begin{cases} \sigma^2\eta/k_j, & j = j', i \neq i' \\ 0, & j \neq j' \end{cases}
\end{aligned}
$$

Hence the pairwise variance for two plots in the same block is $2\sigma^2$, and for two plots in blocks j and j' is $\sigma^2\{2 + \eta(1/k_j + 1/k_{j'})\}$ ($j \neq j'$).

The covariance between two plots in the jth block depends on the block size. As discussed by Patterson and Thompson (1971), this is considered appropriate for field experiments where the larger blocks show the smaller covariance. An alternative would be to specify a model where the covariance was independent of block size. Differences between these options are minimized by ensuring that block sizes do not differ by more than one plot.

Following Williams (1986b) and de Hoog, Speed and Williams (1990), the generalized least squares equations for treatments eliminating replicates can be written as

$$(\mathbf{X}'\mathbf{V}^{*+}\mathbf{X})\hat{\tau} = \mathbf{X}'\mathbf{V}^{*+}\mathbf{y} \tag{7.43}$$

where $\mathbf{V}^* = (\mathbf{I} - \mathbf{P}_R)\mathbf{V}(\mathbf{I} - \mathbf{P}_R)$, $\mathbf{P}_R = \mathbf{R}(\mathbf{R}'\mathbf{R})^{-1}\mathbf{R}'$ is the projection matrix for replicates, and $\mathbf{V}$ is given in (7.42). Since $\hat{\tau}'\mathbf{1} = 0$, it will be necessary to add the overall mean $\bar{y}$ to each of the treatment effects obtained from (7.43) to make them directly comparable with the unadjusted treatment means and the intra-block estimates given by (1.20).

Estimators of σ^2 and η can be obtained by equating the mean squares of suitably chosen quadratic forms to expectation, as was done in Section 7.6. First define the general quadratic form

$$S(\mathbf{Q}) = \mathbf{y}'\{\mathbf{Q} - \mathbf{Q}\mathbf{X}(\mathbf{X}'\mathbf{Q}\mathbf{X})^{+}\mathbf{X}'\mathbf{Q}\}\mathbf{y} \tag{7.44}$$

for some symmetric matrix $\mathbf{Q}$. Then $S(\mathbf{I} - \mathbf{P}_B)$ is the residual sum of squares in the intra-block analysis of variance of a resolvable block design, so that RMS $= S(\mathbf{I} - \mathbf{P}_B)/f_B$ is the residual mean square with $f_B = vr - v - rs + 1$ degrees of freedom and where s is the number of blocks per replicate. RMS provides an estimator of σ^2. A complete block analysis of variance would have residual mean square CMS $= S(\mathbf{I} - \mathbf{P}_R)/f_R$ where $f_R = (r-1)(v-1)$. Then an estimator for η is provided by

$$\hat{\eta} = \frac{(v-1)(\text{CMS} - \text{RMS})}{(s-1)\text{RMS}} \tag{7.45}$$

For equal block sizes (7.45) is equivalent to the estimator $\hat{\eta} = k\hat{\sigma}_1^2/\hat{\sigma}^2$, where $\hat{\sigma}^2$ and $\hat{\sigma}_1^2$ are given by (7.40); this can be demonstrated as follows. From the analysis of variance tables for the intra-block and complete block models $r(s-1)\text{BMS} + f_B\text{RMS} = f_R\text{CMS}$, where BMS is the adjusted block mean square. Then using (7.40)

$$\begin{aligned}
\hat{\eta} &= \frac{r(\text{BMS} - \text{RMS})}{(r-1)\text{RMS}} \\
&= \frac{f_R\text{CMS} - \{f_B + r(s-1)\}\text{RMS}}{(r-1)(s-1)\text{RMS}}
\end{aligned}$$

which simplifies to (7.45).

When $\hat{\eta}$ is negative, i.e. RMS $>$ CMS, it is usual to set η to zero in (7.42) so that the model then reverts to that for a complete block design.

The REML method can also be used to estimate σ^2 and η. Note that both GENSTAT and SAS assume that plot variances are independent of block sizes; see the discussion earlier in this section. However, the variance matrix in (7.42) can be specified in the latest versions of these packages. In any case when block sizes differ by at most one plot, variation in estimates resulting from different models should be minimal.

Non-resolvable incomplete block designs with unequal block sizes are rarely used. However for this situation the model (7.42) can be used by excluding the term for replicates, namely π. Then by equating $\text{PMS} = S(\mathbf{I})/f_P$ to expectation, where $f_P = v(r - 1)$, it can be shown that

$$\hat{\eta} = \frac{rv(\text{PMS} - \text{RMS})}{b\text{RMS}}$$

7.8 Neighbour analysis of field trials

The model (7.42) for a resolvable incomplete block design has proved very valuable for field trials. For example, Patterson and Hunter (1983) analysed 244 cereal trials and reported an average efficiency of 1.43 compared with complete block analysis. However for plots that are side by side, this model has been criticized because it makes only partial allowance for relationships between neighbouring plots. In general too little weight is given to differences between neighbouring plots and too much to differences between plots that are further apart than average. To try and overcome this problem Bartlett (1978) investigated Papadakis's (1937) idea of adjusting yields by covariance on the residuals of neighbouring plots. Wilkinson, Eckert, Hancock and Mayo (1983) then proposed a method called NN analysis in which local trends are removed by taking second differences of the data.

Williams (1986b) approached the problem by proposing a neighbour-type refinement to the incomplete block model (7.42), namely

$$\mathbf{V} = \sigma^2(\mathbf{I} + \eta\mathbf{P}_B - \phi\mathbf{F}) \qquad (7.46)$$

where $\mathbf{F}$ is block diagonal with components $3\mathbf{L}_{k_j}/(k_j^2 - 1)$ for the jth block, and $\mathbf{L}_k$ is a $k \times k$ matrix with (i,j)th element given by $|i - j|$. The pairwise variance between two plots in the jth block and separated by $d-1$ other plots is therefore $2\sigma^2\{1+3\phi d/(k_j^2-1)\}$ while for two plots in different blocks this variance remains the same as in the incomplete block model. Thus, within a block, the pairwise variance is linearly related to the distance between plots. This model is called the *linear variance plus incomplete block* (LV+IB) model. The scaling factors $3/(k_j^2 - 1)$ in $\mathbf{F}$ act as normalizing constants in a manner similar to $1/k_j$ in $\mathbf{P}_B$; for example it can be shown that $\mathbf{P}_B \mathbf{F} \mathbf{P}_B = \mathbf{P}_B$.

The variance matrix (7.46) reduces to that for the incomplete block model when ϕ is set to zero. On the other hand if incomplete blocks are ignored then (7.46) becomes $\mathbf{V} = \mathbf{I} - \phi \mathbf{F}$ with $s = 1$, i.e. $k = v$. This is called the *linear variance* (LV) model. Besag and Kempton (1986) also proposed a neighbour model based on first differences of the data and noted that this is equivalent to the LV model.

Williams (1986b) estimated σ^2, η and ϕ by equating quadratic forms such as $S(\mathbf{I} - \mathbf{P}_R)$ and $S(\mathbf{I} - \mathbf{P}_B)$ to expectation, whereas Gleeson and Cullis (1987) and Baird and Mead (1991) used REML to estimate σ^2 and η in the LV model.

If $\hat{\phi}$ is negative for a particular trial then by analogy with what is done for incomplete blocks, ϕ is set to zero and the model reverts to (7.42) or to the complete block model.

Example 7.1

The example in Section 7.6.2 consists of a line of 72 plots and so is suitable for either LV+IB or LV analysis. For the LV+IB model the variance estimates using the method of Williams (1986b) are $\hat{\sigma}^2 = 0.0426$, $\hat{\eta} = 9.351$ and $\hat{\phi} = 2.872$, whereas for the LV model $\hat{\sigma}^2 = 0.0517$ and $\hat{\phi} = 37.98$. The efficiencies relative to a complete block design are 1.42 and 1.79 respectively and so the LV model is more appropriate for this example. Typically the LV+IB model is more successful when block sizes are much larger than the $k = 4$ here. Estimated treatment means for the two models are given in Table 7.6, where for convenience the combined means from Table 7.5 are repeated.

Although the results of neighbour analyses appear impressive relative to incomplete block analyses, the extra complexity of the models and estimation procedures mean that careful screening of results is usually required which perhaps limits the general use of these methods. Also very little analytical work has been done to investigate possible bias in the estimation of σ^2.

The above discussion on neighbour methods has been based on

Table 7.6. *Adjusted variety means*

Variety	Combined estimates	LV+IB estimates	LV estimates
1	5.11	5.10	5.01
5	5.04	5.08	5.05
15	4.97	5.00	4.96
19	4.84	4.82	4.79
21	4.80	4.76	4.76
14	4.77	4.78	4.83
13	4.76	4.83	4.78
12	4.76	4.71	4.78
16	4.73	4.75	4.75
17	4.60	4.61	4.63
6	4.54	4.49	4.52
22	4.53	4.48	4.53
8	4.53	4.51	4.56
4	4.49	4.42	4.54
2	4.48	4.47	4.43
10	4.37	4.32	4.28
18	4.36	4.37	4.39
11	4.28	4.33	4.36
23	4.25	4.15	4.18
24	4.15	4.22	4.26
7	4.11	4.08	4.05
20	4.04	4.11	4.11
9	3.50	3.54	3.45
3	3.50	3.61	3.54
SED	0.265	0.252	0.224

analysing plot-to-plot variation in one direction. However, plots are often laid out in a two-dimensional array and Cullis and Gleeson (1991) have developed a two-dimensional neighbour model appropriate for this situation. Kempton, Seraphin and Sword (1994) have carried out an extensive investigation of one- and two-dimensional neighbour analyses of data from 224 cereal trials and compared the results with incomplete block and row–column analyses. Interestingly the efficiencies for row–column analyses are comparable with those obtained using two-dimensional neighbour analyses, namely 1.53 and 1.59 respectively. This difference would have been reduced if efficient row–column designs such as those produced by ALPHA+ had been used in every case.

CHAPTER 8

Factorial experiments: single and fractional replication

8.1 Introduction

Previous chapters have been concerned with experiments in which a single set of treatments was applied to the plots in a block design or in a row–column design. Suppose now that there are two different treatments, say temperature and pressure, and that their effect on the response of some chemical process is to be studied. One approach would be to carry out two separate experiments. One would be concerned with the effect produced by varying temperature levels, with pressure kept at some constant value. The other experiment would look at the effect produced by varying pressure levels with temperature now held constant.

An alternative approach would be to study the effect of temperature and pressure simultaneously. An experiment could be carried out in which different combinations of the two treatments, temperature and pressure, are used and their effects assessed. Such an experiment is an example of a *factorial experiment*. The effect of temperature can now be obtained either at each different level of pressure or averaged over the levels of pressure. In a similar way, the effects of pressure can be examined. Additionally, a factorial experiment provides information on how the two treatments *interact* with each other. It may be, for instance, that the effect of temperature is different at the different levels of pressure. Such information is not available when carrying out separate experiments. Only by changing the experimental conditions in this way is it possible to examine not only the effects of temperature and pressure separately but also the way they interact.

As a further example, consider an experiment to compare the weight increases of animals kept on different diets using animals of different breeds fed by different methods. Using a factorial experiment it will be possible to determine the best diets, breeds and methods. Also it will be possible to study whether, for instance, the best method of feeding varies from diet to diet or whether the optimum method and diet depend upon the breed of the animal.

Each basic treatment, such as temperature or diet, will be called a *factor*. The number of possible forms of a factor will be called the number of *levels* of the factor. For example, if there are three diets then the factor diets will be at three levels. A particular combination of factors determines a treatment or *treatment combination*. A *symmetrical factorial* experiment or design will have all factors at the same number of levels, otherwise it will be *asymmetrical*. The effect of a factor averaged over the levels of all the other factors is called the *main effect* of that factor. The way in which the effects of factors change at different levels of other factors represents the *interaction* between factors.

Treatment combinations can be set out in any of the designs given in earlier chapters. However, attention will be restricted primarily to factorial experiments in block designs; the use of row–column designs will be discussed in Section 9.8. Again the allocation of treatment combinations to plots should be aimed at maximizing the amount of information on treatments from comparisons made within blocks. With complete block designs all treatment contrasts are estimated entirely within blocks, but constraints on block size often make the use of such designs impractical. If the number of treatment combinations is not too large then it is often possible to set out the experiment in incomplete block designs in which some information is available within blocks on all factorial effects (main effects and interactions); such effects are said to be *partially confounded* with blocks. However, factorial experiments with many factors, or with factors at many levels, involve a large number of treatment combinations. The use of designs which require a number of replicates of each treatment combination then becomes impractical. For instance, an experiment with four factors each at three levels involves 81 treatment combinations so that a block design with only two replicates requires taking 162 observations.

To overcome this problem, designs using just a single replicate are frequently employed. Information on all or part of some of the factorial effects will consequently no longer be available from comparisons within blocks; these effects, or some components of them, will be said to be *totally confounded* with blocks. A further problem is that, since there are no treatment replications, an estimate of error has to be obtained by assuming that certain effects, usually the high order interactions, are negligible. Even single replicate designs are sometimes too large to use in practice so that fractional designs have to be used; these are discussed in Section 8.10.

The principle of confounding will now be illustrated by considering in detail a factorial experiment with two factors F_1 and F_2 each at three levels, the 3^2 factorial experiment. Before doing so, however, main effects and interaction in the two-factor experiment will be defined.

More generally, let the two factors F_1 and F_2 be at m_1 and m_2 levels respectively. A treatment combination will be denoted by the 2-tuple a_1a_2, where a_i represents a level of factor F_i ($a_i = 0, 1, \ldots, m_i - 1$; $i = 1, 2$). For the 3^2 experiment the nine treatment combinations are, therefore, given by

$$00 \quad 01 \quad 02 \quad 10 \quad 11 \quad 12 \quad 20 \quad 21 \quad 22$$

Suppose that all m_1m_2 treatment combinations are accommodated in a single block, and let y_{ij} be the observation obtained from applying treatment combination ij. Often it will be more convenient to represent y_{ij} simply by ij, letting ij stand for both the treatment combination and its corresponding observation. Let

$$\bar{y}_{i.} = \sum_j y_{ij}/m_2, \quad \bar{y}_{.j} = \sum_i y_{ij}/m_1, \quad \bar{y} = \sum_i \sum_j y_{ij}/m_1m_2$$

The effect of factor F_1 at level i is measured by

$$F_{1i} = \bar{y}_{i.} - \bar{y} \quad (i = 0, 1, \ldots, m_1 - 1) \tag{8.1}$$

and is the comparison of the mean of the observations having factor F_1 at level i with the overall mean. Alternatively, it can be regarded as the deviations $y_{ij} - \bar{y}_{.j}$ averaged over the levels of factor F_2. The effects $F_{10}, F_{11}, \ldots$ are (treatment) contrasts in the observations and measure the *main effect* of factor F_1. It is a measure of the extent to which levels of F_1 differ when averaged over the levels of factor F_2. Note that these contrasts are not linearly independent since they sum to zero. There are in fact $(m_1 - 1)$ linearly independent contrasts, so that the main effect of F_1 is based on $(m_1 - 1)$ degrees of freedom.

Similarly, the main effect of factor F_2, with $(m_2 - 1)$ degrees of freedom, is defined in terms of

$$F_{2j} = \bar{y}_{.j} - \bar{y} \quad (j = 0, 1, \ldots, m_2 - 1) \tag{8.2}$$

and again measures the extent to which levels of factor F_2 differ when averaged over the levels of factor F_1.

Suppose now that the effect of factor F_1 at level i is the same when measured at each level of factor F_2, i.e.

$$F_{1i} = y_{ij} - \bar{y}_{.j}, \quad \text{for all } j \qquad (8.3)$$

If (8.3) is true for all levels of factor F_1 then the two factors F_1 and F_2 are said to be *additive*, since the difference between any two observations y_{ij} and $y_{i'j'}$ is then given by the sum of differences in their (main) effects, i.e.

$$\begin{aligned} y_{ij} - y_{i'j'} &= (\bar{y}_{i.} - \bar{y}_{i'.}) + (\bar{y}_{.j} - \bar{y}_{.j'}) \\ &= (F_{1i} - F_{1i'}) + (F_{2j} - F_{2j'}) \end{aligned}$$

If two factors are not additive then they are said to *interact*. The interaction between the two factors F_1 and F_2, the $F_1 F_2$ interaction, is then measured by the set of contrasts $(y_{ij} - \bar{y}_{.j}) - F_{1i}$, i.e.

$$(F_1 F_2)_{ij} = y_{ij} - \bar{y}_{i.} - \bar{y}_{.j} + \bar{y} \quad \text{for all } i, j \qquad (8.4)$$

Factors F_1 and F_2 are interchangeable in (8.3) so that the contrasts (8.4) are also given by $(y_{ij} - \bar{y}_{i.}) - F_{2j}$. The interaction contrasts (8.4) are not linearly independent and, since $\sum_i (F_1 F_2)_{ij} = \sum_j (F_1 F_2)_{ij} = 0$, are based on $(m_1 - 1)(m_2 - 1)$ degrees of freedom.

Finally, an identity which will be useful in the examples that follow is

$$y_{ij} = F_{1i} + F_{2j} + (F_1 F_2)_{ij} + \bar{y} \qquad (8.5)$$

Returning now to the 3^2 experiment, if all nine treatment combinations can be accommodated in a single block then both main effects and the interaction can be estimated within blocks. Consider, however, the following single replicate design for the 3^2 experiment in three blocks of three, where as usual blocks are written in columns:

$$\begin{array}{ccc} 00 & 01 & 02 \\ 11 & 12 & 10 \\ 22 & 20 & 21 \end{array}$$

Certain treatment contrasts are no longer estimable within blocks since the design is disconnected. For instance, there are two linearly independent treatment contrasts involving block totals which are not estimable. Thus, letting B_1, B_2 and B_3 represent the three block totals, the block contrast $B_1 - B_3$ corresponds to a contrast in the treatment combinations, namely $(00+11+22)-(02+10+21)$. This treatment contrast is said to be totally confounded with the block contrast. Another treatment contrast is confounded with $B_1 - 2B_2 + B_3$, a block contrast orthogonal to $B_1 - B_3$. Hence, two independent treatment contrasts are totally confounded with blocks. It now remains to show that the two confounded treatment contrasts correspond, in this design, to two degrees of freedom of the $F_1 F_2$ interaction.

It is clear from the design that main effects are estimable. Every level of a factor occurs once in each block so that main effect contrasts can be estimated independently of the block totals. For instance,

$$F_{10} - F_{11} = \tfrac{1}{3}\{(00 - 11) + (01 - 12) + (02 - 10)\}$$

involving three differences, each of which is estimable within blocks. It follows that components of the $F_1 F_2$ interaction must be confounded with blocks. This can be demonstrated more rigorously by showing that the block contrasts can be expressed as linear combinations of the interaction contrasts $(F_1 F_2)_{ij}$ given in (8.4). Using (8.5) it can be verified that, writing f_{ij} for $(F_1 F_2)_{ij}$,

$$B_1 - B_3 = (f_{00} + f_{11} + f_{22}) - (f_{02} + f_{10} + f_{21})$$

and

$$B_1 - 2B_2 + B_3 = (f_{00} + f_{11} + f_{22}) - 2(f_{01} + f_{12} + f_{20}) + (f_{02} + f_{10} + f_{21})$$

It can also be noted that the three treatment combinations in the first block of the design satisfy the relationship $a_1 + 2a_2 = 0$ modulo 3, where again a_i denotes a level of the ith factor $(i = 1, 2)$. Further, the treatments in the second block satisfy $a_1 + 2a_2 = 1$ modulo 3, and those in the third block $a_1 + 2a_2 = 2$ modulo 3. The two degrees of freedom component of the $F_1 F_2$ interaction involving contrasts in the three sets of treatment combinations satisfying the equations

$$a_1 + 2a_2 = 0, 1, 2 \quad (\text{modulo } 3)$$

will be denoted by $F_1 F_2^2$; the powers attached to the letters are the coefficients in the linear combination of a_1 and a_2.

The remaining two degrees of freedom of the $F_1 F_2$ interaction, which are necessarily estimable within blocks, correspond to the $F_1 F_2$ component of the interaction, i.e. to contrasts in the sets of treatment combinations satisfying the equations

$$a_1 + a_2 = 0, 1, 2 \quad (\text{modulo } 3)$$

The two components $F_1 F_2$ and $F_1 F_2^2$ represent an orthogonal decomposition of the F_1F_2 interaction. The four linearly independent contrasts defined by these components span the interaction space given by the contrasts (8.4).

Not all single replicate designs will lead to components of certain main effects and interactions being totally confounded with blocks. Consider, for instance, the following alternative design for the 3^2 experiment in three blocks of three:

$$\begin{array}{ccc} 00 & 12 & 02 \\ 01 & 20 & 10 \\ 22 & 21 & 11 \end{array}$$

It follows from (8.5) that, for instance

$$B_1 - B_3 = (2F_{10} + F_{12} + f_{00} + f_{01} + f_{22})$$
$$- (F_{10} + 2F_{11} + f_{02} + f_{10} + f_{11})$$

so that part of the main effect of F_1 and part of the F_1F_2 interaction are confounded with blocks.

Attention is restricted to block designs typified by the first example, in which each treatment contrast confounded with blocks belongs entirely within a particular main effect or interaction.

8.2 Treatment structure

Consider again a factorial experiment with two factors F_1 and F_2 at m_1 and m_2 levels respectively. If these treatment combinations are set out in a block design then the appropriate model is given by (1.3), namely

$$y_{ikl} = \mu + \tau_i + \beta_k + \varepsilon_{ikl}$$

where, in particular, τ_i is the effect of the ith treatment combination. It will be more convenient in a two-factor experiment to write the treatment effect as τ_{ij}, with the two suffixes identifying the levels of the two factors. Thus, τ_{ij} is the effect of the treatment combination having factors F_1 at the ith level and F_2 at the jth level. Now consider the following identity

$$\tau_{ij} = \bar{\tau} + (\bar{\tau}_{i.} - \bar{\tau}) + (\bar{\tau}_{.j} - \bar{\tau}) + (\tau_{ij} - \bar{\tau}_{i.} - \bar{\tau}_{j.} + \bar{\tau}) \qquad (8.6)$$

where $\bar{\tau}_{i.}$ and $\bar{\tau}_{.j}$ are means averaged over the levels of F_2 and F_1 respectively, and $\bar{\tau}$ is the overall mean of the τ_{ij}. When compared with (8.1) and (8.2), it is clear that the second and third terms on the right-hand side of (8.6) represent main effects of F_1 and F_2 respectively, while from a comparison with (8.4), the final term represents the $F_1 F_2$ interaction. Hence, (8.6) gives a partition of the $m_1 m_2 - 1$ linearly independent treatment contrasts $\tau_{ij} - \bar{\tau}$ into three (orthogonal) components representing the main effects and interaction based on $m_1 - 1$, $m_2 - 1$ and $(m_1 - 1)(m_2 - 1)$ degrees of freedom respectively. This partition defines the treatment structure of the factorial experiment.

In matrix notation, the set of equations given by (8.6) can be written as

$$\tau = \mathbf{C}_{00}\tau + \mathbf{C}_{10}\tau + \mathbf{C}_{01}\tau + \mathbf{C}_{11}\tau \qquad (8.7)$$

where

$$
\begin{aligned}
\mathbf{C}_{00} &= \mathbf{K}_{m_1} \otimes \mathbf{K}_{m_2} \\
\mathbf{C}_{10} &= (\mathbf{I}_{m_1} - \mathbf{K}_{m_1}) \otimes \mathbf{K}_{m_2} \\
\mathbf{C}_{01} &= \mathbf{K}_{m_1} \otimes (\mathbf{I}_{m_2} - \mathbf{K}_{m_2}) \\
\mathbf{C}_{11} &= (\mathbf{I}_{m_1} - \mathbf{K}_{m_1}) \otimes (\mathbf{I}_{m_2} - \mathbf{K}_{m_2})
\end{aligned}
$$

where $\mathbf{K}_{m_j}$ is an $m_j \times m_j$ matrix with every element equal to m_j^{-1}. Premultiplying a vector by the matrix $\mathbf{K}$ results in each element of the vector being replaced by its mean, whereas premultiplying by $\mathbf{I} - \mathbf{K}$ results in the mean being subtracted from each element in the vector. Hence, premultiplying τ by $\mathbf{C}_{00}$ can be seen as an operation which first averages over the levels of F_2 and then over the levels of F_1 to give a vector with each element equal to $\bar{\tau}$. Premultiplying τ by $\mathbf{C}_{10}$ first averages over the levels of F_2 and then subtracts the overall mean from each resulting mean. Similarly, premultiplying τ by $\mathbf{C}_{01}$ results in the overall mean being subtracted from means obtained by averaging over the levels F_1. The product $\mathbf{C}_{11}\tau$ can best be understood by looking at the two parts of the operation. First, premultiplying by $\mathbf{I} \otimes (\mathbf{I} - \mathbf{K})$ produces a vector of differences $\tau_{ij} - \bar{\tau}_{i.}$ where averaging is over the levels of F_2. Premultiplying by $\mathbf{K} \otimes (\mathbf{I} - \mathbf{K})$ now results in these differences being averaged over the levels of F_1 to give differences $\bar{\tau}_{.j} - \bar{\tau}$. Putting the two operations together gives a vector whose elements are of the form

$$(\tau_{ij} - \bar{\tau}_{i.}) - (\bar{\tau}_{.j} - \bar{\tau}) = \tau_{ij} - \bar{\tau}_{i.} - \bar{\tau}_{.j} + \bar{\tau}$$

which is the required contrast.

A concise representation of the $\mathbf{C}$ matrices in (8.7) is given by

$$\mathbf{C}_{x_1 x_2} = \mathbf{C}_{x_1} \otimes \mathbf{C}_{x_2}$$

where

$$\mathbf{C}_{x_j} = \begin{cases} \mathbf{K}_{m_j}, & x_j = 0 \\ \mathbf{I}_{m_j} - \mathbf{K}_{m_j}, & x_j = 1 \end{cases} \qquad (8.8)$$

For three factors F_1, F_2 and F_3 at m_1, m_2 and m_3 levels respectively the vector of treatment parameters τ can be partitioned into eight components given by

$$\tau = \sum \mathbf{C}_{x_1 x_2 x_3} \tau \qquad (x_i = 0, 1; \; i = 1, 2, 3)$$

where

$$\mathbf{C}_{x_1 x_2 x_3} = \mathbf{C}_{x_1} \otimes \mathbf{C}_{x_2} \otimes \mathbf{C}_{x_3}$$

and where $\mathbf{C}_{x_j}$ is given by (8.8). The terms in the summation correspond to, respectively, the mean effect, the main effects of F_1, F_2 and F_3, the two-factor interactions $F_1 F_2$, $F_1 F_3$ and $F_2 F_3$ and, additionally, the three-factor interaction $F_1 F_2 F_3$. This last interaction measures the extent to which the $F_1 F_2$ interaction varies at different levels of F_3 or, equivalently, the $F_1 F_3$ interaction different levels of F_2 or the $F_2 F_3$ interaction at different levels of F_1. A negligible $F_1 F_2 F_3$ interaction means that the interaction of any two of the factors is the same at each level of the third factor.

For an n-factor experiment with factor F_i at m_i levels ($i = 1, 2, \ldots, n$) the treatment structure is given by

$$\tau = \sum_x \mathbf{C}_x \tau \tag{8.9}$$

where the summation is over all binary numbers $x = (x_1 x_2 \ldots x_n)$ and

$$\mathbf{C}_x = \mathbf{C}_{x_1} \otimes \mathbf{C}_{x_2} \otimes \ldots \otimes \mathbf{C}_{x_n} = \bigotimes_{j=1}^{n} \mathbf{C}_{x_j} \tag{8.10}$$

and where $\mathbf{C}_{x_j}$ is given by (8.8). A *generalized interaction* (i.e. main effect or interaction) is represented by $\mathbf{C}_x \tau$, where factor F_i is present in the interaction if $x_i = 1$ ($i = 1, 2, \ldots, n$). This generalized interaction will be denoted by $F^x = F_1^{x_1} F_2^{x_2} \ldots F_n^{x_n}$. For example, in a five-factor experiment the $F_1 F_2 F_4$ interaction is represented by the set of contrasts $\mathbf{C}_{11010} \tau$.

These $\mathbf{C}_x$ matrices form a *complete binary set* in that they are symmetric ($\mathbf{C}_x = \mathbf{C}'_x$), idempotent ($\mathbf{C}_x^2 = \mathbf{C}_x$), orthogonal ($\mathbf{C}_x \mathbf{C}_y = \mathbf{0}, x \neq y$) and satisfy $\sum \mathbf{C}_x = \mathbf{I}$. These results follow as both $\mathbf{K}$ and $(\mathbf{I} - \mathbf{K})$ are symmetric idempotent matrices, as $\mathbf{K}(\mathbf{I} - \mathbf{K}) = \mathbf{0}$, and by noting that

$$\sum_x \mathbf{C}_x = \sum \bigotimes_{j=1}^{n-1} \mathbf{C}_{x_j} \otimes \{\mathbf{K} + (\mathbf{I} - \mathbf{K})\}$$

where the second summation is over all binary numbers $(x_1 x_2 \ldots$

x_{n-1}). The result $\sum \mathbf{C}_x = \mathbf{I}$, implied by (8.9), can then be verified by induction.

Note also that all $\mathbf{C}_x$ matrices, apart from $\mathbf{C}_0$, are contrast matrices. This follows from the orthogonality property, as then $\mathbf{C}_x \mathbf{C}_0 = \mathbf{0}$ implies $\mathbf{C}_x \mathbf{1} = \mathbf{0}$ for all $x \neq 0$. Finally, since $\mathbf{C}_x$ is idempotent,

$$\mathrm{rank}(\mathbf{C}_x) = \mathrm{trace}(\mathbf{C}_x) = \prod_{j=1}^{n} \mathrm{trace}(\mathbf{C}_{x_j})$$

and the trace of $\mathbf{C}_{x_j}$ is 1 if $x_j = 0$, and $m_j - 1$ if $x_j = 1$. Hence

$$\mathrm{rank}(\mathbf{C}_x) = \prod_{j=1}^{n} (m_j - 1)^{x_j} \qquad (8.11)$$

which gives the degrees of freedom associated with the generalized interaction $\mathbf{C}_x \tau$.

Although (8.9) represents the treatment structure of an n-factor experiment, not all generalized interactions need be included in the fitted model. It is often the case in practice that only a subset of these interactions is considered for inclusion. In single replicate designs some effects, often the higher order interactions, are usually assumed to be negligible and are used to provide an estimate of experimental error.

8.3 Complete block designs

Suppose that the $v = m_1 m_2 \dots m_n$ treatment combinations of an n-factor experiment having factor F_i at m_i levels ($i = 1, 2, \dots, n$) are set out in a complete block design with $b = r$ blocks. The design consisting of a single block accommodating all treatment combinations is a special case with $r = 1$. To avoid duplication, results will now be given which apply when $r \geq 1$, although the use of multiple replication designs will be discussed in Chapter 9.

The analysis of a complete block design is given in Sections 1.1 and 1.9. The covariances between the estimators $\mathbf{C}_x \hat{\tau}$ and $\mathbf{C}_y \hat{\tau}$ of two different generalized interactions are

$$\mathrm{cov}(\mathbf{C}_x \hat{\tau}, \mathbf{C}_y \hat{\tau}) = \mathbf{C}_x \mathbf{\Omega} \mathbf{C}_y \sigma^2 = \mathbf{C}_x \mathbf{C}_y (\sigma^2/r) = \mathbf{0} \qquad (8.12)$$

for all $x \neq y$, since from (1.49) $\mathbf{\Omega} = (1/r)\mathbf{I}$. Hence, the estimate of any contrast in the factorial effect $\mathbf{C}_x \tau$ is orthogonal to the estimate of every contrast in the effect $\mathbf{C}_y \tau$. Factorial effects are then said to be orthogonal to each other. This means that any factorial effect can be estimated independently of the other effects and without regard to which other effects have been included in the model. The adjusted treatment sum of squares can, in consequence, be partitioned orthogonally into sums of squares corresponding to each of the factorial effects.

From results in Section 1.9, the treatment contrast $\mathbf{C}_x\tau$ is estimable with estimator

$$\mathbf{C}_x\hat{\tau} = (1/r)\mathbf{C}_x\mathbf{T} \qquad (8.13)$$

where $\mathbf{T}$ is the vector of treatment totals. Thus estimates are given by contrasts in the unadjusted treatment means. The variance–covariance matrix of this estimator is given by

$$V(\mathbf{C}_x\hat{\tau}) = \mathbf{C}_x\Omega\mathbf{C}_x\sigma^2 = \mathbf{C}_x(\sigma^2/r) \qquad (8.14)$$

The sum of squares due to testing the hypothesis $H_0 : \mathbf{C}_x\tau = \mathbf{0}$ is, from (1.38),

$$S(H_0) = (\mathbf{C}_x\hat{\tau})'(\mathbf{C}_x\Omega\mathbf{C}_x)^-(\mathbf{C}_x\hat{\tau})$$

i.e., using (8.13) and (8.14) and with $\mathbf{C}_x^- = \mathbf{I}$,

$$S(H_0) = (1/r)(\mathbf{C}_x\mathbf{T})'(\mathbf{C}_x\mathbf{T}) \qquad (8.15)$$

This sum of squares is based on ν degrees of freedom, where $\nu = \text{rank}(\mathbf{C}_x)$ is given by (8.11).

The calculation of the sum of squares in (8.15) is particularly straightforward. To illustrate this, consider a factorial experiment with three factors F_1, F_2 and F_3 at m_1, m_2 and m_3 levels respectively. Let the treatment total corresponding to the treatment combination with factor F_j at level i_j be denoted by $T_{i_1i_2i_3}$ $(0 \le i_j \le m_j - 1; j = 1, 2, 3)$ and let

$$\bar{T}_{i_1i_2.} = \sum_{i_3} T_{i_1i_2i_3}/m_3, \quad \bar{T}_{i_1..} = \sum_{i_2}\sum_{i_3} T_{i_1i_2i_3}/(m_2m_3)$$

and so on. For the main effect of factor F_1,

$$\mathbf{C}_{100}\mathbf{T} = \{(\mathbf{I} - \mathbf{K}) \otimes \mathbf{K} \otimes \mathbf{K}\}\mathbf{T} = \bar{\mathbf{T}}_{F_1} - \bar{T}_{...}\mathbf{1}$$

where $\bar{\mathbf{T}}_{F_1}$ is a vector consisting of m_2m_3 replications of each of the m_1 elements $\bar{T}_{i_1...}$. Hence, the sum of squares due to the main effect of factor F_1 is

$$S(F_1) = (m_2m_3/r) \sum_{i_1=0}^{m_1-1} (\bar{T}_{i_1..} - \bar{T}_{...})^2$$

$$= \sum_{i_1=0}^{m_1-1} T_{i_1..}^2/(rm_2 m_3) - \text{CF} \qquad (8.16)$$

where $\text{CF} = T_{...}^2/(rm_1 m_2 m_3)$ is the correction factor for the overall mean. Hence, $S(F_1)$ is equal to the sum of squares of factor F_1 totals divided by the number of observations making up each of these totals less the overall correction factor. It is the same type of formula as that obtained from fitting a set of parameters in a one-way analysis of variance model; see (1.32) or (1.35).

To obtain the $F_1 F_2$ interaction sum of squares $S(F_1 F_2)$ let

$$S(F_1 \text{ and } F_2) = \sum_{i_1} \sum_{i_2} T_{i_1 i_2.}^2/(rm_3) - \text{CF}$$

Then, since

$$\begin{aligned}
\mathbf{C}_{110} &= (\mathbf{I} - \mathbf{C}_{00} - \mathbf{C}_{10} - \mathbf{C}_{01}) \otimes \mathbf{K} \\
&= (\mathbf{I} \otimes \mathbf{I} \otimes \mathbf{K}) - (\mathbf{K} \otimes \mathbf{K} \otimes \mathbf{K}) - \mathbf{C}_{100} - \mathbf{C}_{010}
\end{aligned}$$

it follows that

$$S(F_1 F_2) = S(F_1 \text{ and } F_2) - S(F_1) - S(F_2)$$

The extension to higher factor interactions and to the general factorial experiment should be apparent.

8.4 Identifying components of interaction

For symmetrical p^n designs, where p is a prime or prime power, Bose (1947) and Kempthorne (1947) divided the complete set of treatment degrees of freedom into homogeneous orthogonal subsets; a subset is said to be homogeneous if all its contrasts belong to the same generalized interaction, and two subsets are orthogonal if all contrasts in one are orthogonal to those in the other. For example, it was shown in Section 8.1 that the $F_1 F_2$ interaction in a 3^2 experiment could be divided into two orthogonal components, denoted by $F_1 F_2^2$ and $F_1 F_2$, each based on two degrees of freedom representing contrasts among treatment combinations which satisfied, respectively, the equations

$$\begin{aligned}
a_1 + 2a_2 &= 0, 1, 2 \pmod{3} \\
a_1 + a_2 &= 0, 1, 2 \pmod{3}
\end{aligned}$$

These ideas were extended to include asymmetrical and other symmetrical designs by Bailey, Gilchrist and Patterson (1977).

First, $H(z)$ is defined; it comprises a subset of degrees of freedom associated with the vector $z = z_1 z_2 \ldots z_n$ where z_i is one of the integers $0, 1, \ldots, m_i - 1$. A treatment combination is again represented by $a = a_1 a_2 \ldots a_n$, where a_i is a level of factor F_i $(i = 1, 2, \ldots, n)$. Let the integer $a(z)$ be given by

$$a(z) = \sum_{i=1}^{n} \delta_i a_i z_i \quad (\text{modulo } \gamma) \tag{8.17}$$

where γ is the lowest common multiple of the m_i and where $\delta_i = \gamma / m_i$. Then $H(z)$ is defined to represent the set of all contrasts between treatment combinations with different values of $a(z)$.

For example, for the 3^2 experiment $\gamma = 3$ and $\delta_i = 1$ $(i = 1, 2)$ so that $a(12) = a_1 + 2a_2$ and $a(11) = a_1 + a_2$, modulo 3. Hence, $H(12)$ and $H(11)$ correspond to the $F_1 F_2^2$ and $F_1 F_2$ components respectively of the $F_1 F_2$ interaction.

Consider now a $2 \times 3 \times 6$ experiment, i.e. with three factors F_1, F_2 and F_3 at 2, 3 and 6 levels respectively. Since $\gamma = 6$ and $\delta_1 = 3, \delta_2 = 2$ and $\delta_3 = 1$,

$$a(z) = 3a_1 z_1 + 2a_2 z_2 + a_3 z_3$$

The values of $a(z)$ for $z = 114$ are given in Table 8.1.

There are six different $a(114)$ values so that $H(114)$ represents five degrees of freedom. $H(114)$ is not, however, homogeneous. The contrast between treatment combinations with even and odd values of $a(114)$ is the single F_1 main effect degree of freedom. There are also two degrees of freedom from the interaction $F_2 F_3$; these are given by the contrasts between pairs of values 0 and 3, 1 and 4, 2 and 5. The remaining two degrees of freedom are from the three-factor interaction. Neither are the $H(z)$ orthogonal. For instance,

Table 8.1. *Values of $a(114)$ for $2 \times 3 \times 6$ experiment*

Levels of F_1	Levels of F_2	Levels of F_3					
		0	1	2	3	4	5
0	0	0	4	2	0	4	2
0	1	2	0	4	2	0	4
0	2	4	2	0	4	2	0
1	0	3	1	5	3	1	5
1	1	5	3	1	5	3	1
1	2	1	5	3	1	5	3

$H(114)$ and $H(011)$ can be shown to share the same degrees of freedom from F_2F_3 but otherwise are different, whereas $H(022)$ is a proper subset of $H(114)$.

However, from $H(z)$, sets $H_*(z)$ can be derived that are homogeneous, orthogonal and exhaustive, in the sense that between them they account for all treatment degrees of freedom. In the $2 \times 3 \times 6$ experiment the F_1 main effect in $H(114)$ is identifiable as $H(100)$ and the two degrees of freedom from F_2F_3 are $H(022)$. The two degrees of freedom for the three-factor interaction can therefore be identified as those contrasts in $H(114)$ that are orthogonal to $H(100)$ and $H(022)$. These two degrees of freedom are called $H_*(114)$.

Bailey, Gilchrist and Patterson (1977) give a simple method for obtaining a complete set of $H_*(z)$. The set of all vectors z form an Abelian group G, where addition of two z vectors consists of addition of the individual z_i modulo m_i. Each vector z generates a cyclic subgroup $\langle z \rangle$ given by

$$\langle z \rangle : \quad 0, z, 2z, \ldots, (s-1)z$$

where s is the smallest non-zero integer having $sz = 0$. When the integer r is coprime to s, $\langle rz \rangle = \langle z \rangle$; otherwise $\langle rz \rangle$ is a proper subgroup of $\langle z \rangle$.

$H_*(z)$ is now defined to be the set of all contrasts that are orthogonal to those in $H(s_1z), \ldots, H(s_hz)$, where $s_1, s_2, \ldots, s_h$ are the proper prime divisors of s. The number of degrees of freedom in $H_*(z)$ is the Euler function $\phi(s)$, the number of integers between 1 and s which are coprime to s. Bailey, Gilchrist and Patterson (1977) show that the set of $H_*(z)$ are orthogonal, homogeneous and exhaustive, with the contrasts in $H_*(z)$ belonging to the interaction defined by the non-zero elements of z.

Hence, continuing the $2 \times 3 \times 6$ example,

$$\langle 114 \rangle : \quad 000 \quad 114 \quad 022 \quad 100 \quad 014 \quad 122$$

Now $s = 6$, $\phi(s) = 2$ and $s_1 = 2$, $s_2 = 3$ giving $\langle s_1z \rangle$ and $\langle s_2z \rangle$ as, respectively,

$$\langle 022 \rangle : \quad 000 \quad 022 \quad 014$$
$$\langle 100 \rangle : \quad 000 \quad 100$$

$H_*(114)$, therefore, represents two degrees of freedom from the $F_1F_2F_3$ interaction. It can be shown that the remaining eight degrees of freedom of this interaction correspond to components $H_*(111)$, $H_*(112)$, $H_*(113)$ and $H_*(115)$, each with two degrees of freedom.

For symmetrical p^n experiments, where p is a prime, the $H_*(z)$ components are identical to those given by Bose (1947) and Kempthorne (1947). This follows since $s = p$ and $\phi(s) = p - 1$, so that $H_*(z) = H(z)$ with $p - 1$ degrees of freedom. If p is a prime power, i.e. $p = q^m$ where q is prime, then each treatment factor can be replaced by m pseudo factors each at q levels. The sets of degrees of freedom described by Bose (1947) are then unions of the $H_*(z)$.

8.5 Construction of single replicate designs

In an n-factor experiment with factor F_i at m_i levels a treatment combination is represented by $a = a_1 a_2 \ldots a_n$ where a_i is a level of F_i ($0 \le a_i \le m_i - 1$; $i = 1, 2, \ldots, n$). Addition of two treatment combinations a and b is defined as

$$a + b = a_1 a_2 \ldots a_n + b_1 b_2 \ldots b_n = c_1 c_2 \ldots c_n = c \qquad (8.18)$$

where $c_i = a_i + b_i$, modulo m_i ($i = 1, \ldots, n$). For some non-negative integer u let $ua = a + a + \ldots + a$ (u times), i.e. ua is obtained by multiplying each a_i by u and reducing modulo m_i whenever necessary.

Now in the $2 \times 3 \times 6$ experiment discussed in the previous section, a design confounding the five degrees of freedom in $H(114)$ is obtained by placing the six treatment combinations with $a(114) = 3a_1 + 2a_2 + 4a_3 = 0$, modulo 6, in the initial block; those with $a(114) = 1$, modulo 6, in the next block; and so on. More generally, the treatment combinations in the initial block of a design will satisfy $a(z_i) = 0$, modulo γ, for a set of vectors $z_1, z_2, \ldots$ where $a(z)$ is defined in (8.17). Now if treatment combinations a and b both satisfy this equation then so will treatment combination $u_1 a + u_2 b$, for any non-negative integers u_1 and u_2. This result leads to the following general method of constructing single replicate designs.

Consider the cyclic subgroup of treatment combinations

$$G : 0, a, 2a, \ldots, (q - 1)a \qquad (8.19)$$

where q, the order of the group, is the smallest integer such that $qa = 0$. G is an Abelian group under addition defined by (8.18) and has q distinct elements. Let the q treatment combinations in G comprise the initial block of a design with the remaining blocks given by the *cosets* of G, obtained by adding other treatment combinations to those in G. The resulting design is necessarily a single replicate design. The treatment combination a is the *generator* of G and, hence, of the design.

Example 8.1

Consider a 3^2 experiment with generator $a = 11$. Then $G : 00\ 11\ 22$ and the design is

$$
\begin{array}{ccc}
00 & 01 & 02 \\
11 & 12 & 10 \\
22 & 20 & 21
\end{array}
$$

The design was given earlier in Section 8.1.

More generally, let a_i be a generator of a subgroup G_i of order q_i $(i = 1, 2, \ldots, p)$ then

$$G : G_1 \oplus G_2 \oplus \ldots \oplus G_p \qquad (8.20)$$

is the direct sum of these p subgroups with general element given by

$$u_1 a_1 + u_2 a_2 + \ldots + u_p a_p \qquad (8.21)$$

$(0 \le u_i < q_i;\ i = 1, 2, \ldots, p)$. Assuming that the generators a_i have been chosen in such a way that no treatment combination occurs more than once in G then G is a subgroup of order $q = \prod_{i=1}^{n} q_i$. This can be achieved in a sequential manner by ensuring that a_2 is not in G_1, a_3 is not in $G_1 \oplus G_2$, a_4 is not in $G_1 \oplus G_2 \oplus G_3$, and so on. The q elements of G will then constitute the initial block of a single replicate design with the remaining blocks given by the cosets of G.

Example 8.2

Consider a 4^2 experiment with two generators $a_1 = 02$ and $a_2 = 11$. Then

$$
\begin{array}{lcccc}
G_1 : & 00 & 02 \\
G_2 : & 00 & 11 & 22 & 33
\end{array}
$$

so that

$$G : \quad 00 \quad 11 \quad 22 \quad 33 \quad 02 \quad 13 \quad 20 \quad 31$$

Since $q_1 = 2$ and $q_2 = 4$, the single replicate design obtained using G has two blocks of eight given by

$$
\begin{array}{l}
(00 \quad 02 \quad 11 \quad 13 \quad 20 \quad 22 \quad 31 \quad 33) \\
(01 \quad 03 \quad 12 \quad 10 \quad 21 \quad 23 \quad 32 \quad 30)
\end{array}
$$

Example 8.3

Consider the asymmetrical $3 \times 4 \times 6$ factorial experiment with the two generators $a_1 = 101$ and $a_2 = 023$. Then

$$
\begin{array}{lcccccc}
G_1 : & 000 & 101 & 202 & 003 & 104 & 205 \\
G_2 : & 000 & 023
\end{array}
$$

giving

$$G: \quad \begin{matrix} 000 & 101 & 202 & 003 & 104 & 205 \\ 023 & 124 & 225 & 020 & 121 & 222 \end{matrix}$$

Now $q_1 = 6$ and $q_2 = 2$ so that the resulting single replicate design has six blocks of 12 given by

000	001	002	010	011	012
003	004	005	013	014	015
020	021	022	030	031	032
023	024	025	033	034	035
101	102	103	111	112	113
104	105	100	114	115	110
121	122	123	131	132	133
124	125	120	134	135	130
202	203	204	212	213	214
205	200	201	215	210	211
222	223	224	232	233	234
225	220	221	235	230	231

The choice of appropriate generators will depend on the confounding scheme of the resulting design. The $3 \times 4 \times 6$ design in Example 8.3, for instance, may be of little practical use since one degree of freedom from the main effect of factor F_2 is confounded with blocks; this degree of freedom being the comparison of the first three blocks, having only levels 0 and 2 for the second factor, with the remaining blocks.

Finally, it can be noted that the single replicate designs obtained by the above method belong to the class of n-cyclic designs considered in Section 3.5; they are in fact disconnected partial n-cyclic sets.

8.6 Analysis of single replicate designs

Let the rows of the incidence matrix $\mathbf{N}$ of the single replicate design be permuted so that the first k rows correspond to the k treatment combinations in the first block, the next k rows to the k treatment combinations in the second block, and so on. Then $\mathbf{N} = \mathbf{I}_b \otimes \mathbf{1}_k$ and $\mathbf{NN'} = \mathbf{I}_b \otimes \mathbf{J}_k$, where $b = v/k$ is the number of blocks. Hence, the information matrix $\mathbf{A}$ of (1.14) is given by

$$\mathbf{A} = \mathbf{I} - (1/k)\mathbf{NN'} = \mathbf{I}_b \otimes (\mathbf{I}_k - \mathbf{K}_k) \tag{8.22}$$

Two important results follow directly from the fact that this information matrix is an idempotent matrix.

Firstly, the eigenvalues of $\mathbf{A}$ are either 0 or 1, which means that the canonical efficiency factors of a single replicate design are all equal to unity. In other words, if a treatment contrast is estimable then it is estimated with full efficiency.

Secondly, a generalized inverse of $\mathbf{A}$ is given by $\mathbf{\Omega} = \mathbf{I}$. Hence the analysis of a single replicate design follows along the lines of that of a complete block design, with estimates of generalized interactions given by (8.13) and corresponding sum of squares by (8.15), where $r = 1$ and where the vector of unadjusted treatment totals $\mathbf{T}$ is now replaced by the vector of adjusted treatment totals $\mathbf{q}$; this vector being given by subtracting from each observation (since $r = 1$) its corresponding block mean. These estimates and sums of squares are based on the number of unconfounded degrees of freedom associated with the generalized interactions. If an interaction, as opposed to components of an interaction, is totally confounded then its estimate, sum of squares and degrees of freedom will be zero. If, on the other hand, an interaction is completely unconfounded the analysis proceeds exactly as for a complete block design, since block mean adjustments will cancel each other out.

For confounded main effects and interactions it must be remembered that not all components can be estimated. For instance, consider a four-level factor which has the comparison of levels 0 and 1 with levels 2 and 3 confounded. Now this contrast cannot be estimated free of block effects, so that the main effect sum of squares will not include a contribution from this contrast. If this contrast is likely to be of importance then the design will clearly be unsatisfactory.

It now remains to be able to determine the confounding scheme of any given single replicate design constructed by the methods of Section 8.5.

8.7 Determining the confounding scheme

It has already been stated in Section 8.5 that if a design is required which confounds the subsets of degrees of freedom given by $H(z_1), H(z_2), \ldots$, for some vectors $z_1, z_2, \ldots$, then the initial block will contain those treatment combinations satisfying $a(z_i) = 0$, modulo γ $(i = 1, 2, \ldots)$, where $a(z_i)$ is defined in (8.17). Given a single replicate design, however, the reverse process has to be carried out. The components of interaction confounded are identified by finding z vectors such that the treatment

combinations in the initial block all provide integers defined in
(8.17) equal to zero.

Example 8.4

Consider the following single replicate design for a 6^3 experiment
in six blocks of 36 obtained from the two generators $a_1 = 012$ and
$a_2 = 113$. A vector $z = z_1 z_2 z_3$ must now be found which satisfies
both $a_1(z) = 0$ and $a_2(z) = 0$, i.e.

$$z_2 + 2z_3 = 0 \quad \text{and} \quad z_1 + z_2 + 3z_3 = 0 \quad \text{(modulo 6)}$$

Clearly $z_1 + z_3$ must be zero so that $z_1 = 1$ and $z_3 = 5$ gives $z_2 = 2$.
Now

$$\langle 125 \rangle : \quad 000 \quad 125 \quad 244 \quad 303 \quad 422 \quad 541$$

giving $s = 6$, $\phi(s) = 2$ and

$$\langle 244 \rangle : \quad 000 \quad 244 \quad 422$$
$$\langle 303 \rangle : \quad 000 \quad 303$$

Hence, the confounded components are $H_*(125)$ and $H_*(244)$ each
with two degrees of freedom and $H_*(303)$ with one degree of
freedom.

Example 8.5

Consider the design given in Example 8.3 of a $3 \times 4 \times 6$ experiment
with the two generators $a_1 = 101$ and $a_2 = 023$. The design has
six blocks so that five degrees of freedom will be confounded with
blocks. Now $z = z_1 z_2 z_3$ must satisfy

$$a_1(z) = 4z_1 \quad + 2z_3 = 0 \quad \text{(modulo 12)}$$
$$a_2(z) = \quad\quad 6z_2 + 6z_3 = 0 \quad \text{(modulo 12)}$$

The equations are satisfied by $z = 222$. Now

$$\langle 222 \rangle : \quad 000 \quad 222 \quad 104 \quad 020 \quad 202 \quad 124$$

giving $s = 6$, $s_1 = 2$, $s_2 = 3$, $\phi(s) = 2$ and

$$\langle 104 \rangle : \quad 000 \quad 104 \quad 202$$
$$\langle 020 \rangle : \quad 000 \quad 020$$

Hence, the confounded components are $H_*(222)$ and $H_*(104)$ with
two degrees of freedom and $H_*(020)$ with one degree of freedom.
Thus, components of the main effect of F_2 and the $F_1 F_3$ and
$F_1 F_2 F_3$ interactions have been totally confounded in this design.

In both of the above examples, since there are a small number of blocks and a relatively large block size, the confounded components $H_*(z)$ can be readily found. Consider, however, a 6^3 design in 36 blocks of six obtained from the generator $a = 113$; i.e. with a large number of blocks but small block size. Finding the components $H_*(z)$ now becomes more tedious since there are many z vectors satisfying $z_1 + z_2 + 3z_3 = 0$, modulo 6. For instance, it is satisfied by $z = 301$ so that it can consequently be shown that three degrees of freedom from the F_1F_3 interaction are confounded, which are identified by the components $H_*(301)$ and $H_*(303)$ with two and one degrees of freedom respectively. Although all components can be found in this way, it is sometimes easier to first determine the confounding scheme using the following alternative method.

Consider an n-factor single replicate design constructed according to the methods of Section 8.5. Since these designs are n-cyclic sets, it follows from (3.8) that the concurrence matrix $\mathbf{NN'}$ can be written as

$$\mathbf{NN'} = \sum_{a_1} \sum_{a_2} \cdots \sum_{a_n} \lambda_{a_1 a_2 \ldots a_n} (\mathbf{\Gamma}_{a_1} \otimes \mathbf{\Gamma}_{a_2} \otimes \ldots \otimes \mathbf{\Gamma}_{a_n}) \quad (8.23)$$

where $\mathbf{\Gamma}_{a_i}$ is a basic circulant matrix as defined in Section 3.5 and where $\lambda_{a_1 a_2 \ldots a_n} = 1$ if treatment combination $a_1 a_2 \ldots a_n$ occurs in the initial block and zero otherwise. Now an eigenvector of $\mathbf{NN'}$ is given by

$$\mathbf{h}_x = \mathbf{h}_{x_1} \otimes \mathbf{h}_{x_2} \otimes \ldots \otimes \mathbf{h}_{x_n}$$

where $\mathbf{h}_{x_i}$ is a vector of ones if $x_i = 0$ and is a contrast vector satisfying $\mathbf{h}'_{x_i} \mathbf{1} = 0$ if $x_i = 1$. Further, since $\mathbf{C}_{x_i} \mathbf{h}_{x_i} = \mathbf{h}_{x_i}$ for $x_i = 0$ or 1, where $\mathbf{C}_{x_i}$ is given by (8.8), then $\mathbf{h}_x$ is an eigenvector of $\mathbf{C}_x$ and is therefore in the vector space spanned by the columns of $\mathbf{C}_x$.

Hence there exists a set of contrast vectors which span the generalized interaction space defined by the columns of $\mathbf{C}_x$ and which are also eigenvectors of the information matrix $\mathbf{A}$. The following theorem can now be proved.

Theorem 8.1 *The number of degrees of freedom of the generalized interaction $\mathbf{C}_x\boldsymbol{\tau}$ confounded with blocks is*

$$Y_x = (1/k)\,trace(\mathbf{NN'C}_x) \quad (8.24)$$

Proof. Let the $v \times v$ matrix $\mathbf{C}_x$ have rank $t < v$ and let $\mathbf{H}_x$ be a $v \times t$ matrix of rank t whose columns are eigenvectors of $\mathbf{A}$, i.e. $\mathbf{AH}_x = \mathbf{H}_x \theta^\delta$ where θ^δ is a diagonal matrix with elements $\theta_1, \theta_2, \ldots, \theta_t$. The eigenvalue θ_i is equal to 0 or 1 since

$\mathbf{A}$ is idempotent $(i = 1, 2, \ldots, t)$. Now since $\mathbf{H}'_x \tau$ is estimable if and only if $\mathbf{H}_x = \mathbf{A}\mathbf{H}_x$ (see Section 1.5) then the number of linearly independent estimable contrasts in $\mathbf{H}'_x \tau$ is $\sum_{i=1}^{t} \theta_i$. Let the columns of $\mathbf{H}_x$ be in the vector space spanned by the columns of $\mathbf{C}_x$ so that $\mathbf{H}_x = \mathbf{C}_x \mathbf{H}_x$. Also let $\mathbf{G}_x$ be a $v \times (v - t)$ matrix of rank $v - t$ such that $\mathbf{C}_x \mathbf{G}_x = 0$. It then follows that

$$\mathbf{A}\mathbf{C}_x(\mathbf{H}_x \vdots \mathbf{G}_x) = (\mathbf{H}_x \vdots \mathbf{G}_x) \begin{pmatrix} \theta \\ 0 \end{pmatrix}^{\delta}$$

so that

$$\sum_{i=1}^{t} \theta_i = \text{trace}(\mathbf{A}\mathbf{C}_x)$$

This represents the number of degrees of freedom of the generalized interaction $\mathbf{C}_x \tau$ that are estimable. Hence, the number of degrees of freedom confounded is

$$t - \text{trace}(\mathbf{A}\mathbf{C}_x) = \text{trace}(\mathbf{C}_x) - \text{trace}(\mathbf{A}\mathbf{C}_x) = Y_x$$

which completes the proof.

For the designs constructed by the methods of Section 8.5, (8.24) simplifies to give

$$Y_x = (1/k) \sum (\prod_j w_{a_j}) \tag{8.25}$$

where the summation is over all treatment combinations in the initial block, the product is over all j for which $x_j = 1$ and where

$$w_{a_j} = \begin{cases} m_j - 1 & \text{if } a_j = 0 \\ -1 & \text{if } a_j \neq 0 \end{cases}$$

With $\mathbf{N}\mathbf{N}'$ given by (8.23) it follows that

$$\text{trace}(\mathbf{N}\mathbf{N}'\mathbf{C}_x) = \sum \prod_{j=1}^{n} \text{trace}(\mathbf{\Gamma}_{a_j} \mathbf{C}_{x_j})$$

If $x_j = 0$ then $\mathbf{\Gamma}_{a_j} \mathbf{C}_{x_j} = \mathbf{K}$ so that its trace is 1. If $x_j = 1$ then $\mathbf{\Gamma}_{a_j} \mathbf{C}_{x_j} = \mathbf{\Gamma}_{a_j} - \mathbf{K}$ and

$$\text{trace}(\mathbf{\Gamma}_{a_j} - \mathbf{K}) = \begin{cases} m_j - 1 & \text{if } a_j = 0 \\ -1 & \text{if } a_j \neq 0 \end{cases}$$

The result given in (8.25) is thus established.

A number of examples will now be given to illustrate how Y_x in (8.25) is calculated. Results can be conveniently set out in tabular form with the treatment combinations in the initial block in the first column and the w_{a_j} and their products in the other columns.

Table 8.2. *Calculations for confounding in the 3^2 experiment*

Initial block $a_1 a_2$	w_{a_1}	w_{a_2}	$w_{a_1} w_{a_2}$
00	2	2	4
11	−1	−1	1
22	−1	−1	1
Total	0	0	6

Example 8.6

For the 3^2 experiment of Example 8.1 with generator $a = 11$ the calculations are set out in Table 8.2. Hence

$$Y_{10} = \tfrac{1}{3} \sum w_{a_1} = 0, \qquad Y_{01} = \tfrac{1}{3} \sum w_{a_2} = 0,$$
$$Y_{11} = \tfrac{1}{3} \sum w_{a_1} w_{a_2} = 2$$

confirming that two degrees of freedom from the $F_1 F_2$ interaction are confounded with blocks.

Example 8.7

For the single replicate design of Example 8.3 for a $3 \times 4 \times 6$ experiment with generators $a_1 = 101$ and $a_2 = 023$ the calculations are given in Table 8.3. These confirm that one degree of freedom from the F_2 main effect and two degrees of freedom from the $F_1 F_3$ and $F_1 F_2 F_3$ interactions are confounded with blocks.

Example 8.8

Returning to the 6^3 experiment in 36 blocks of six obtained from the generator $a = 113$, the results are given in Table 8.4.

The use of (8.25) gives a simple way of obtaining the number of degrees of freedom confounded with blocks for any generalized interaction. It does not identify these degrees of freedom with the components $H_*(z)$ of the interaction but, if it is necessary to specify these components, this can be done subsequently. This task is in any case simplified when the general confounding scheme is known, especially when there are a large number of blocks.

When the number of levels of a factor is not a prime number then the use of *pseudo factors* can often be used to enlarge the class of available designs. More specifically, Voss and Dean (1987)

Table 8.3. *Calculations for confounding in the* $3 \times 4 \times 6$ *experiment*

Initial block	w_{a_1}	w_{a_2}	w_{a_3}	$w_{a_1}w_{a_2}$	$w_{a_1}w_{a_3}$	$w_{a_2}w_{a_3}$	$w_{a_1}w_{a_2}w_{a_3}$
000	2	3	5	6	10	15	30
003	2	3	-1	6	-2	-3	-6
020	2	-1	5	-2	10	-5	-10
023	2	-1	-1	-2	-2	1	2
101	-1	3	-1	-3	1	-3	3
104	-1	3	-1	-3	1	-3	3
121	-1	-1	-1	1	1	1	-1
124	-1	-1	-1	1	1	1	-1
202	-1	3	-1	-3	1	-3	3
205	-1	3	-1	-3	1	-3	3
222	-1	-1	-1	1	1	1	-1
225	-1	-1	-1	1	1	1	-1
Total	0	12	0	0	24	0	24
Y_x	0	1	0	0	2	0	2

Table 8.4. *Calculations for confounding in the* 6^3 *experiment*

Initial block	w_{a_1}	w_{a_2}	w_{a_3}	$w_{a_1}w_{a_2}$	$w_{a_1}w_{a_3}$	$w_{a_2}w_{a_3}$	$w_{a_1}w_{a_2}w_{a_3}$
000	5	5	5	25	25	25	125
113	-1	-1	-1	1	1	1	-1
220	-1	-1	5	1	-5	-5	5
333	-1	-1	-1	1	1	1	-1
440	-1	-1	5	1	-5	-5	5
553	-1	-1	-1	1	1	1	-1
Total	0	0	12	30	18	18	132
Y_x	0	0	2	5	3	3	22

have shown that when the number of levels of a factor has a prime powered divisor then different designs may sometimes be obtained by the use of pseudo factors. The number of levels associated with each pseudo factor will be a prime number. Suppose, for instance, that a factor F_i at four levels is represented by two pseudo factors F_{i1} and F_{i2} each at two levels. The correspondence between the levels of F_i and F_{i1} and F_{i2} will be as shown in Table 8.5. The three degrees of freedom of the main effect of F_i correspond to the

Table 8.5. *Correspondence between levels of factor and pseudo factors*

Factor	Level			
F_i	0	1	2	3
F_{i1}	0	0	1	1
F_{i2}	0	1	0	1

main effects F_{i1} and F_{i2} and the interaction $F_{i1}F_{i2}$, each with a single degree of freedom. The degrees of freedom of the interaction of factor F_i with factor F_j ($j \neq i$) will be identified with those of the interactions $F_{i1}F_j$, $F_{i2}F_j$ and $F_{i1}F_{i2}F_j$. Other interactions will be given in a similar way.

Using the pseudo factors, designs can now be constructed by the methods of Section 8.5. The confounding scheme of the pseudo factor design can be determined by the above methods. Although care has to be taken in identifying the interactions confounded in the pseudo factor design with those in the corresponding design for the original factors, the procedure presents no real difficulties. An example will illustrate the technique.

Example 8.9

Consider a factorial experiment with four factors F_1, F_2, F_3 and F_4 at 2, 3, 4 and 6 levels respectively. Suppose a single replicate design in 12 blocks of 12 is required. Let factor F_3 be represented by two pseudo factors F_{31} and F_{32} each at two levels. For the $2 \times 3 \times (2 \times 2) \times 6$ experiment, the two generators $a_1 = 11101$ and $a_2 = 00013$ produce a design of the required size, which confounds one degree of freedom from F_1F_{31}, $F_{31}F_{32}F_4$ and $F_1F_{32}F_4$, and two degrees of freedom from F_2F_4, $F_2F_{31}F_{32}F_4$, $F_1F_2F_{31}F_4$ and $F_1F_2F_{32}F_4$, i.e. one degree of freedom from F_1F_3, F_3F_4 and $F_1F_3F_4$, two degrees of freedom from F_2F_4 and $F_2F_3F_4$ and four degrees of freedom from $F_1F_2F_3F_4$. If pseudo factors are not used then it is not possible to construct a design with more than two degrees of freedom from the four-factor interaction $F_1F_2F_3F_4$ confounded with blocks.

Finally, it should be noted that, given the generators, it is a relatively straightforward task to write a computer program which will determine the confounding scheme of a single replicate design using the above methods. It is particularly simple to write a program to calculate Y_x given in (8.25). First, the initial block of the design is obtained from (8.19) and (8.20) by constructing the cyclic subgroup G_i for each generator a_i ($i = 1, 2, \ldots, p$) and then taking the direct sum of these subgroups. Next, the quantities w_{a_j} can be calculated for each factor in turn for every treatment combination in the initial block. Finally, the number of degrees of freedom confounded with blocks for each generalized interaction can be calculated from Y_x.

Determining which components of the generalized interactions are confounded can also be programmed without too much difficulty. The main steps in the program are as follows:

1. Obtain the set z of all z-vectors satisfying $a_i(z) = 0$ for $i = 1, 2, \ldots, p$.

2. Find the order of each cyclic subgroup generated from every element in z.

3. From z take a z-vector, z_1 say, corresponding to a group of lowest order. Then $H_*(z_1)$ is confounded, with degrees of freedom given by the number of non-zero elements in the group. Now delete all z-vectors belonging to this group from z.

4. From the remaining elements in z, take a z-vector corresponding to a group of lowest order. This gives another confounded component with degrees of freedom given by the number of non-zero z-vectors in this group which are still in z. All elements belonging to this group are then deleted from z. The procedure is repeated until all components have been found.

For instance, Example 8.4 would give

$$z\ :\quad 125\quad 244\quad 303\quad 422\quad 541$$
$$\text{Group order}:\quad 6\quad\ \ 3\quad\ \ 2\quad\ \ 3\quad\ \ 6$$

and the algorithm would produce the confounding scheme given in the example.

The availability of such a computer algorithm together with a careful choice of generators would enable a design of the required size and with the required properties, assuming such a design exists, to be quickly found.

8.8 Choice of generators

The size of the experiment and the required confounding scheme determine the choice of generators for the single replicate design.

Appropriate generators can usually be found quickly by trial and error, as the confounding pattern of any given set of generators is easily obtained. Of course, the constraints imposed by the size and required confounding scheme may be such that no suitable design can be found.

Simple rules for choosing generators can be derived from (8.25) to ensure that all main effects are unconfounded, which is usually an essential requirement. For instance, suppose a design is to be constructed using a single generator $a = a_1 a_2 \ldots a_n$ where the ith factor F_i is at m_i levels ($i = 1, 2, \ldots, n$). Then the main effect F_i is unconfounded if and only if a_i is relatively prime to m_i, i.e. $(m_i, a_i) = 1$. In this case

$$g_i : \quad 0 \quad a_i \quad 2a_i \quad \ldots \quad (m_i - 1)a_i \qquad (8.26)$$

constitutes a group closed to addition modulo m_i. Each zero must, therefore, be accompanied in the initial block by $m_i - 1$ non-zero elements. Hence, the sum $\sum w_{a_i}$ in (8.25) is zero.

With more than one generator it is only necessary for m_i to be relatively prime to the ith element in one of the generators for the F_i main effect to be unconfounded. Again this follows from (8.26) and (8.25). For instance, with two generators $a = a_1 a_2 \ldots a_n$ and $b = b_1 b_2 \ldots b_n$ the F_i main effect will be unconfounded if and only if $(m_i, a_i, b_i) = 1$. Consider, as an example, the $3 \times 4 \times 6$ experiment with generators $a_1 = 101$ and $a_2 = 023$ given in Example 8.3. The F_2 main effect is confounded since $(4,0,2) = 2$, but the F_1 and F_3 main effects are not since $(3,1,0) = 1$ and $(6,1,3) = 1$.

For a two-factor interaction to be unconfounded it is necessary that, for the treatment combinations in the initial block, each distinct level of one factor should be accompanied by every level of the other factor. For example, the design for a 4^2 experiment with generators $a_1 = 21$ and $a_2 = 22$ has initial block

$$00 \quad 21 \quad 02 \quad 23 \quad 22 \quad 03 \quad 20 \quad 01$$

The four treatment combinations with factor F_1 at level 0 have factor F_2 at each of its four levels; similarly for F_1 at level 2. Hence, the $F_1 F_2$ interaction will be unconfounded. Again, for the $2 \times 3 \times 4$ design with generator $a = 111$ it can be readily verified from the 12 treatment combinations in the initial block that $F_1 F_2$ and $F_2 F_3$ interactions are unconfounded.

One particular case when the $F_i F_j$ interaction, say, is necessarily unconfounded is where at least one of the generators has its ith element zero and jth element non-zero and relatively prime to m_j, or vice-versa. For instance, in the $3 \times 4 \times 6$ design of Example 8.3 inspection of the two generators immediately shows that $F_1 F_2$ and $F_2 F_3$ are unconfounded. For a further example, a 4^3 experiment in 16 blocks of four can be obtained using two generators both of order 4. Using the above results, it can be readily established that use of generators $a_1 = 101$ and $a_2 = 011$ will ensure that all main effects and all two-factor interactions are unconfounded.

For three-factor interactions each distinct pair of levels of two of the factors must be accompanied by every level of the third factor. The extension to any interaction is obvious, although direct application of the methods in Section 8.7 would probably be more appropriate for interactions involving more than two factors.

8.9 2^n experiments

2^n factorial experiments involving n factors each at two levels are widely used in practice and have been extensively studied. A rigorous theory for constructing 2^n designs corresponding to particular confounding schemes is available and is given in most books on the design of experiments; see, for instance, John and Quenouille (1977, Chapter 5). This theory is a special case of that presented here for single replicate designs.

For a 2^n experiment a treatment combination $a = a_1 a_2 \ldots a_n$ will have $a_i = 0$ or 1 for all i, corresponding to, say, the absence or low level and presence or high level respectively of the factor. Each generator $a \, (a \neq 0)$ will, therefore, give rise to a group G consisting of just two elements 0 and a. The initial or *principal* block of a 2^n design consists of 0 and the elements of the direct sum of the generators. For example, suppose there are three distinct generators a, b and c then the initial block will be

$$0, \ a, \ b, \ a+b, \ c, \ a+c, \ b+c, \ a+b+c$$

To ensure that no treatment combination occurs more than once in the initial block, $c \neq a + b$.

The conventional notation for a treatment combination in a 2^n experiment is to represent the high level of, say, factor A by its lower-case letter a and the low level by the absence of that letter, with treatment combination $00 \ldots 0$ denoted by (1). For example, with four factors $A, \ B, \ C, \ D$,

$$1010 = ac, \quad 0111 = bcd, \quad 0010 = c, \quad 0000 = (1)$$

The addition of two treatment combinations $a + b$ corresponds to multiplication in the conventional notation with the modulo 2 rule that $a^2 = b^2 = c^2 = d^2 = 1$. For instance,

$$1010 + 0111 = 1101 \quad \text{or} \quad ac \times bcd = abc^2d = abd$$

From the results of Section 8.4 the integer $a(z) = \sum a_i z_i$ (modulo 2) can take one of two values, namely 0 or 1. Hence, the component $H(z)$ comprises one degree of freedom and corresponds to the generalized interaction $\mathbf{C}_x\boldsymbol{\tau}$ where $x_i = z_i$ for all i. Hence, $\mathbf{C}_x\boldsymbol{\tau}$ is confounded with blocks if and only if for all treatment combinations in the principal block

$$\sum_{i=1}^{n} a_i x_i = 0 \quad \text{(modulo 2)} \tag{8.27}$$

In fact, it is only necessary for (8.27) to hold for the generators of the principal block since if $\sum a_i x_i = 0$ and $\sum b_i x_i = 0$ then $\sum (a_i + b_i)x_i = 0$, all modulo 2.

The condition given by (8.27) is the 'even–odd' rule usually used for choosing the principal block. Consider, for instance, a 2^4 experiment with factors A, B, C, D and suppose ABC is to be confounded. Then (8.27) becomes $a_1 + a_2 + a_3 = 0$, modulo 2. Hence, an even or zero number of the a_i $(i = 1, 2, 3)$ must be equal to 1. Using the conventional notation, each generator must have an even (or zero) number of letters in common with ABC. Possible generators are, for example, $0001, 1101$ or 1010, i.e. d, abd and ac.

Similar results follow if more than one effect is to be confounded. It follows from (8.27) that if $\mathbf{C}_x\boldsymbol{\tau}$ and $\mathbf{C}_y\boldsymbol{\tau}$ $(x \neq y \neq 0)$ are confounded with blocks then so will $\mathbf{C}_{x+y}\boldsymbol{\tau}$. For example, suppose ABC and ACD are confounded in the 2^4 experiment; then $x = 1110$ and $y = 1011$ so that $x + y = 0101$, showing that BD is also confounded. Again addition of effects corresponds to the multiplication of effects using the conventional notation with the rule that $A^2 = B^2 = C^2 = \ldots = 1$. Hence, $ABC \times ACD = A^2BC^2D = BD$. The treatment combinations in the principal block will have an even or zero number of letters in common with ABC and ACD (and also BD). The set of generalized interactions confounded together with the identity I are called the *defining contrasts* of the 2^n design and form a group closed to either addition or multiplication modulo 2 depending on whether the notation of this chapter or the conventional 2^n notation is used.

8.10 Fractional replication

The number of treatment combinations in a factorial experiment increases rapidly with the number of factors tested. Even with a single replicate design, the size of the experiment will frequently be too large to carry out in practice. If certain factorial effects, usually the high-order interactions, can be assumed to be negligible then it may be possible to use designs that require only a fraction of the total number of treatment combinations and which still permit the important effects, main effects and two-factor interactions say, to be estimated. Such designs are called *fractional* factorial designs.

For example, consider a 3×2^5 experiment involving 96 treatment combinations. Suppose only main effects and two-factor interactions are of interest, with all other effects assumed to be negligible. Testing all 96 treatment combinations now becomes unnecessary since by taking a suitable $\frac{1}{2}$-replicate design, involving 48 treatment combinations, it will be possible to estimate all main effects and two-factor interactions and still provide 21 degrees of freedom for error.

The fractional factorial designs considered here will consist of the treatment combinations in the initial block of the single replicate designs constructed by the methods of Section 8.5.

Consider the single replicate design for three factors F_1, F_2 and F_3 at 2, 3 and 4 levels respectively in two blocks of 12 obtained from the generator $a = 111$. The single degree of freedom confounded with blocks is from the $F_1 F_3$ interaction, being in fact the component $H_*(102)$. A $\frac{1}{2}$-replicate design is obtained by taking the treatment combinations comprising the initial block of this single replicate design. The fractional factorial design consists, therefore, of the 12 treatment combinations

000 002 010 012 020 022 101 103 111 113 121 123

Since $H_*(102)$ is estimated in the single replicate design by the difference between the two block totals, then this component is completely lost in the fractional design. Now consider the single degree of freedom for the F_1 main effect, namely the $H_*(100)$ component. It represents the comparison between the first six treatment combinations in the design, having factor F_1 at level 0, with the remaining six treatment combinations having F_1 at level 1. However, this particular comparison also represents one degree of freedom from the F_3 main effect, the $H_*(002)$ component. It can be seen that the first six treatment combinations have factor F_3 at levels 0 and 2 while the remaining six have F_3 at levels 1 and 3. That is, $H_*(100)$ and $H_*(002)$ are estimated by the same contrast. These two components are then said to be *aliased*. It can be shown that the remaining two degrees of freedom of the F_3 main effect, $H_*(001)$, are aliased with two degrees of freedom from the F_1F_3 interaction, the $H_*(101)$ component. Thus, $H_*(001)$ and $H_*(101)$ are aliased. The aliases of the other components can also be found and, hence, the full aliasing system of the design determined.

An effect is estimable only if its aliases can be assumed to be negligible. To assess the usefulness of any fractional design it is, therefore, important that the aliasing system can be readily obtained.

Assume that in the single replicate design $H(z)$ is confounded with blocks. For each treatment combination a in the initial block the integer $a(z)$ defined by (8.17) will, therefore, equal zero. $H(z)$ is a defining contrast of the design. For $x \neq y \neq z$, $H_*(x)$ and $H_*(y)$ will be aliased if and only if they are estimated by the same contrasts, i.e. if and only if, for each treatment combination a in the initial block, $a(x) = a(y)$. In fact, it is only necessary that $a(x) = a(y)$ for the generating treatment combinations since if the equality holds for treatment combinations a and b it must also hold for $a + b$. For the $2 \times 3 \times 4$ example above,

$$a(z) = 6a_1z_1 + 4a_2z_2 + 3a_3z_3 \quad \text{(modulo 12)}$$

so with defining contrast $H(102)$,

$$a(102) = 6a_1 + 6a_3 \quad \text{(modulo 12)}$$

All 12 treatment combinations given above have $a(102) = 0$ modulo 12. Also, since $a(100) = a(002)$ for the generator $a = 111$, $H_*(100)$ is aliased with $H_*(002)$ and, as $a(001) = a(103)$, $H_*(001)$ is aliased with $H_*(103) = H_*(101)$. The remaining aliases can be established in this way.

Consider the components $H_*(x)$ and $H_*(y)$ and suppose $y = x+t$ for some vector t, where addition of two vectors is defined in Section 8.4. Then, from (8.17)

$$a(y) = \sum_{i=1}^{n} \delta_i a_i(x_i + t_i) = a(x) + a(t) \quad \text{(modulo } \gamma) \quad (8.28)$$

It follows that $a(y) = a(x)$ if and only if $a(t) = 0$, i.e. if and only if t is some multiple of the vector z. The aliases of $H_*(x)$ can, therefore, be determined from the subgroup given by the direct sum of the cyclic subgroups $\langle x \rangle$ and $\langle z \rangle$. Some examples will illustrate the procedure.

Example 8.10
Consider a 3^4 experiment in which $H(1111)$ is the defining contrast of a single replicate design in three blocks of 27. The corresponding factorial design will be a $\frac{1}{3}$-replicate with all treatment combinations satisfying $a_1 + a_2 + a_3 + a_4 = 0$, modulo 3. The 27 treatment combinations can be obtained from the three generators 1110, 0111 and 1011. Now since

$$\langle 1000 \rangle + \langle 1111 \rangle : \quad \begin{array}{ccc} 0000 & 1111 & 2222 \\ 1000 & 2111 & 0222 \\ 2000 & 0111 & 1222 \end{array}$$

it follows that $H_*(1000)$ is aliased with $H_*(2111) = H_*(1222)$ and with $H_*(0222) = H_*(0111)$. The notation

$$H_*(1000) \equiv H_*(1222) \equiv H_*(0111)$$

will be used to show that they are aliased. Alternatively, using the notation of Section 8.1, the F_1 main effect is aliased with components $F_1 F_2^2 F_3^2 F_4^2$ and $F_2 F_3 F_4$. It can be shown that all main effects are aliased with three- and four-factor interactions so that main effects are estimable if these interactions can be assumed to be negligible. However, some components of two-factor interactions are aliased with components of other two-factor interactions. For instance,

$$H_*(1100) \equiv H_*(0011) \equiv H_*(1122)$$

Other two-factor components are aliased with interactions involving three or more factors, for example

$$H_*(1200) \equiv H_*(0122) \equiv H_*(1022)$$

This $\frac{1}{3}$-replicate will provide a useful design if three- and four-factor interactions and some two-factor interactions can be assumed to be negligible. Alternatively, and this applies to all fractional factorials, the design could be used as an initial screening experiment in which the factorial effects likely to be of importance are identified, with another smaller experiment subsequently carried out to separate out any important aliased effects.

If the number of treatment combinations in a fractional factorial experiment is large then it may be necessary to set them out in a number of blocks. For instance, in the 3^4 factorial experiment of Example 8.10 the 27 treatment combinations could be set out in three blocks of nine as follows. Suppose that the two degrees of freedom of the $H_*(1022)$ component of the $F_1 F_3 F_4$ interaction are confounded with blocks. Of course, since $H_*(1022)$ is aliased with $H_*(1200)$ and $H_*(0122)$ both of these components will also be confounded. The three blocks are obtained by arranging the 27 treatment combinations in the $\frac{1}{3}$-replicate to satisfy the equations $a(1022) = 0, 1, 2$, modulo 3. The generators of the initial block can be taken as 1110 and 1101, both of which satisfy $a(1111) = 0$ and $a(1022) = 0$, modulo 3. The full design is

```
(0000  1110  2220  1101  2211  0021  2202  0012  1122)
(0111  1221  2001  1212  2022  0102  2010  0120  1200)
(0222  1002  2112  1020  2100  0210  2121  0201  1011)
```

Setting out a fractional factorial design in blocks, therefore, presents no additional problems apart from the extra care needed in choosing the components to be confounded, since any components aliased with these chosen components will also be confounded.

Example 8.11
Now consider the $3 \times 4 \times 6$ factorial design in six blocks of 12 with generators $a_1 = 101$ and $a_2 = 023$ given in Example 8.3. The five degrees of freedom confounded with blocks was shown in Example 8.5 to be the subset $H(222)$, which was identified with components $H_*(222)$ and $H_*(104)$ with two degrees of freedom each and $H_*(020)$ with one degree of freedom. A $\frac{1}{6}$-replicate, with defining contrast $H(222)$, is given by the 12 treatment combinations in the initial block of this single replicate design, namely

$$000 \quad 003 \quad 020 \quad 023 \quad 101 \quad 104 \quad 121 \quad 124 \quad 202 \quad 205 \quad 222 \quad 225$$

Now the aliases of the F_1 main effect can be determined from the direct sum of the cyclic subgroups $\langle 100 \rangle$ and $\langle 222 \rangle$, which is

$$
\begin{array}{cccccccc}
\langle 100 \rangle + \langle 222 \rangle : & 000 & 222 & 104 & 020 & 202 & 124 \\
& 100 & 022 & 204 & 120 & 002 & 224 \\
& 200 & 122 & 004 & 220 & 102 & 024
\end{array}
$$

Hence

$$H_*(100) \equiv H_*(002) \equiv H_*(022) \equiv H_*(120) \equiv H_*(102) \equiv H_*(122)$$

In a similar way it can be established that the $H_*(010)$ component of the F_2 main effect is aliased with the $H_*(114)$ component of the

$F_1 F_2 F_3$ interaction. However, although the component $H_*(010)$ represents two degrees of freedom from the F_2 main effect, it is clear that only one degree of freedom is estimable in this fractional replicate; each treatment combination has the second factor at either level 0 or level 2. This degree of freedom is thus aliased with one of the four degrees of freedom of the $H_*(114)$ component. In general, the number of degrees of freedom of the $H_*(z)$ component which will be estimable in the fraction is determined by the number of distinct values given by the function $a(z)$ for the treatment combinations in the fraction. For this example,

$$a(z) = 4a_1 z_1 + 3a_2 z_2 + 2a_3 z_3 \quad (\text{modulo } 12)$$

It can be verified that $a(010)$ takes values 0 or 6 for the above 12 treatment combinations, showing that only one degree of freedom of this main effect component is estimable.

Aliases of other effects can be similarly established using the direct sum of cyclic subgroups approach and the number of estimable degrees of freedom determined from the function $a(z)$. To set out the 12 treatment combinations in two blocks of six the single degree of freedom from $H_*(010) \equiv H_*(114)$ could be chosen to be confounded with blocks. As an alternative, the single degree of freedom from $H_*(013) \equiv H_*(111)$ could be used. This would give the design

$$
\begin{array}{cccccc}
(000 & 023 & 104 & 121 & 202 & 225) \\
(003 & 020 & 101 & 124 & 205 & 222)
\end{array}
$$

Note that $a(013) = 0$ for all treatment combinations in the initial or principal block and $a(013) = 6$ for those in the other block, where all quantities are reduced modulo 12 where necessary.

Of course, these $\frac{1}{6}$-replicate designs in either a single block or in two blocks of six have little practical appeal since main effect components are either used in the defining contrasts, or are aliased with each other or are not estimable.

In a 2^n experiment, the above results show that if $H(z_i)$ is a defining contrast of the fractional factorial design then $H(x)$ is aliased with $H(x + z_i)$ for all i. Using the conventional notation, aliases are readily established from the defining contrasts by use of the multiplication rule of Section 8.9. For example, in a 2^6 experiment with factors A, B, C, D, E and F, a $\frac{1}{4}$-replicate can be obtained using the defining contrasts

$$I = ABCD = ABEF = CDEF$$

so that
$$A \equiv BCD \equiv BEF \equiv ACDEF$$
$$AB \equiv CD \equiv EF \equiv ABCDEF$$
and so on, using on multiplication $A^2 = B^2 = \ldots = F^2 = 1$. If all interactions involving three or more factors are assumed negligible, then all main effects are estimable although some two-factor interactions will still be aliased with other two-factor interactions.

In Section 8.7 the outline of a computer program to find the confounding scheme of a single replicate design given its generators was described. Again, using the methods of this section, it is not difficult to add to this program to determine the aliasing scheme of the fractional design resulting from using only the initial block of the confounded design.

The analysis of fractional factorial designs follows along similar lines to that given in Section 8.6 for single replicate designs. Any component of a generalized interaction confounded in the single replicate design will now be completely lost in the fraction. The estimators and sums of squares of other components are obtained as in Section 8.6, where these quantities will necessarily be equal for any two components that are aliased.

8.11 Other designs

A variety of methods have been given for the construction of single replicate and fractional designs for factorial experiments. The earliest method was given by Bose and Kishen (1940) and by Bose (1947) and was applicable to symmetric p^n experiments, where p is a prime number. The method was later generalized by White and Hultquist (1965) to include experiments of the form $p^m q^n$, where p and q are distinct primes. Other generalizations were given by Raktoe (1969), Worthley and Banerjee (1974) and Sihota and Banerjee (1981). Cotter (1974) presented a method for constructing s^n designs, where s is not restricted to being a prime. All of these methods have, however, been shown by Voss and Dean (1987) to give designs which are identical to those produced by the method of Section 8.5.

Patterson (1965, 1976) described an algorithm, incorporated in a computer program called DSIGN, to construct designs and identify confounding patterns for factorial experiments. The algorithm can be used to obtain single replicate block designs in which every component $H_*(z)$ is either totally confounded, partially confounded or unconfounded. Fractional factorial block designs and designs using more complex blocking structures, such as row–column designs, can also be constructed. Bailey (1977) restricted attention from the very broad class of designs generated by the DSIGN algorithm to the subclass in which every component $H_*(z)$ is either completely confounded or is unconfounded. The method of construction given in Section 8.5 follows that given by Bailey (1977), although again the results presented therein are more general than those given here.

Finally, John and Dean (1975) and Dean and John (1975) have given a method of obtaining single replicate designs using the n-cyclic method of construction. As has already been stated, the designs produced by the method of Section 8.5 are in fact disconnected partial n-cyclic sets, and are thus identical to those given by Dean and John.

Single replicate and fractional designs having factors at two and three levels have been extensively tabulated; see, for instance, the volumes produced by the National Bureau of Standards (1957, 1959, 1961) and reproduced in McLean and Anderson (1984); also Greenfield (1976, 1978), Box, Hunter and Hunter (1978), Fries and Hunter (1980) and Franklin (1984). Other designs have been tabulated by Dean and John (1975) and Lewis (1982).

Factorial experiments can also be set out in single and fractional replicate row–column designs; the way this can be done is considered in Section 9.8.

CHAPTER 9

Factorial experiments: multiple replication

9.1 Introduction

More precise estimates of treatment contrasts can be obtained if treatment combinations are replicated more than once in the factorial design. If the number of treatment combinations is not too large or if a large amount of experimental material is available then multiple replicate factorial designs can often be employed. A further advantage is that an independent estimate of experimental error will be available from a comparison among replicate treatments. Such designs will be discussed in this chapter.

One method of constructing multiple replicate factorial designs is to take replications of the single replicate designs considered in the previous chapter. For instance, suppose a design is required for a 3^2 experiment in blocks of three with each treatment combination replicated twice. Two replicates of the design confounding component $F_1 F_2^2$ of the $F_1 F_2$ interaction could be used. No information will be available within blocks on this component but all other effects will be estimable free of block effects. An alternative would be to use single replicate designs which confound different components of the factorial effects; in this example, for instance, the $F_1 F_2^2$ component could be confounded in one replicate and the $F_1 F_2$ component in the second replicate. Now full information is still available on both main effects, and both components of the interaction can be estimated within blocks from half the observations. This interaction is said to be *partially confounded* with blocks. Such a design is frequently to be preferred since some information is available within blocks on all treatment contrasts.

The 3^2 design with two replicates confounding respectively $F_1 F_2^2$

and $F_1 F_2$ is obtained from the two initial blocks (00 11 22) and (00 12 21). The six blocks are

$$
\begin{array}{cccccc}
00 & 01 & 02 & 00 & 01 & 02 \\
11 & 12 & 10 & 12 & 10 & 11 \\
22 & 20 & 21 & 21 & 22 & 20
\end{array}
$$

This design belongs to the class of 2-cyclic designs defined in Section 3.5. The two distinct canonical efficiency factors of the design can be shown to be 1 and $\frac{1}{2}$, each with multiplicity 4. The first of these factors is associated with treatment contrasts corresponding to the two main effects, which are estimated with full information; while the second factor corresponds to interaction contrasts, which are estimable within blocks from half the observations.

In general, obtaining multiple replicate designs from the single replicate designs constructed by the method of Section 8.5 is too restrictive. Suppose, for instance, a design for a 3×2^2 experiment in blocks of four is required. The only single replicate design of this size would confound the main effect of the three-level factor, so that taking replications of this design would clearly be unsatisfactory. Again, for a 4^2 experiment in blocks of six, no single replicate design can be constructed. The approach adopted in this chapter is to set out the treatment combinations of the factorial experiment in incomplete blocks, using primarily the designs of Chapter 3, and to study those designs, in particular, whose canonical efficiency factors can be identified with main effect and interaction treatment contrasts. Designs based on single replicate designs are, as the above 3^2 example shows, included in this general framework.

The use of complete block designs in factorial experiments has already been considered in Section 8.3. The analysis of such designs is straightforward as each treatment contrast is estimable entirely within blocks. Estimates are, therefore, based on unadjusted treatment means, see (8.13). These designs will not be discussed further here.

Results in Chapter 1 on the analysis of incomplete block designs and results on the treatment structure of factorial experiments given in Section 8.2 will be used. In particular, for some n-digit binary number $x = (x_1 x_2 \ldots x_n)$, contrasts in the $F_1^{x_1} F_2^{x_2} \ldots F_n^{x_n}$ interaction are estimated by

$$
\mathbf{C}_x \hat{\boldsymbol{\tau}} = \mathbf{C}_x \boldsymbol{\Omega} \mathbf{q} \tag{9.1}
$$

where $\mathbf{C}_x$ is given by (8.10), $\mathbf{q}$ is the vector of adjusted treatment totals given by (1.15) and $\boldsymbol{\Omega}$ is any generalized inverse of the information matrix $\mathbf{A}$ given by (1.14). Further,

$$\mathrm{cov}(\mathbf{C}_x\hat{\boldsymbol{\tau}}, \mathbf{C}_y\hat{\boldsymbol{\tau}}) = \mathbf{C}_x\boldsymbol{\Omega}\mathbf{C}_y\sigma^2 \qquad (9.2)$$

9.2 Balanced incomplete blocks

Suppose that the v treatment combinations of a factorial experiment are set out in a balanced incomplete block design.

Example 9.1
A balanced incomplete block design for 4^2 treatment combinations in 16 blocks of six units per block is given by

```
00 01 02 03 10 11 12 13 20 21 22 23 30 31 32 33
01 02 03 00 11 12 13 10 21 22 23 20 31 32 33 30
02 03 00 01 12 13 10 11 22 23 20 21 32 33 30 31
10 11 12 13 20 21 22 23 30 31 32 33 00 01 02 03
23 20 21 22 33 30 31 32 03 00 01 02 13 10 11 12
30 31 32 33 00 01 02 03 10 11 12 13 20 21 22 23
```

It can be verified that each treatment combination is replicated six times and that every pair of treatment combinations occurs together in two blocks. The parameters of this design are, therefore, $v = 4^2 = 16$, $r = k = 6$, $b = 16$ and $\lambda = 2$.

No problems of analysis arise since every treatment contrast is a basic contrast in a balanced incomplete block design, with all canonical efficiency factors equal to $E = \lambda v/rk$; see Section 1.7 and (2.12). The covariance matrix of the estimators of two different generalized interactions $\mathbf{C}_x\boldsymbol{\tau}$ and $\mathbf{C}_y\boldsymbol{\tau}$ $(x \neq y)$ is

$$\mathrm{cov}(\mathbf{C}_x\hat{\boldsymbol{\tau}}, \mathbf{C}_y\hat{\boldsymbol{\tau}}) = \mathbf{C}_x\boldsymbol{\Omega}\mathbf{C}_y\sigma^2 = (k/\lambda v)\mathbf{C}_x\mathbf{C}_y\sigma^2 = \mathbf{0} \qquad (9.3)$$

since $\boldsymbol{\Omega} = (k/\lambda v)\mathbf{I}$ is a generalized inverse of the information matrix of a balanced incomplete block design, given in (1.39).

Hence, balanced incomplete block designs share with complete block designs the important property that, for factorial experiments, the estimate of any contrast in one generalized interaction is orthogonal to the estimate of every contrast in any other generalized interaction. Every factorial effect can, thus, be estimated independently of other effects and the adjusted treatment sum of squares can consequently be partitioned orthogonally into sums of squares for each of the factorial effects.

In a balanced incomplete block design, the generalized interaction $\mathbf{C}_x\boldsymbol{\tau}$ is estimated by

$$\mathbf{C}_x\hat{\boldsymbol{\tau}} = (rE)^{-1}\mathbf{C}_x\mathbf{q} \tag{9.4}$$

i.e. by contrasts in the adjusted means of the treatment combinations. The variance–covariance matrix of this estimator is

$$V(\mathbf{C}_x\hat{\boldsymbol{\tau}}) = (rE)^{-1}\mathbf{C}_x\sigma^2 \tag{9.5}$$

and the sum of squares due to testing the hypothesis $H_0 : \mathbf{C}_x\boldsymbol{\tau} = \mathbf{0}$ is

$$S(H_0) = (rE)^{-1}(\mathbf{C}_x\mathbf{q})'(\mathbf{C}_x\mathbf{q}) \tag{9.6}$$

These results are similar in form to those for complete block designs given in Section 8.3. In particular, apart from the factor E^{-1}, the sum of squares in (9.6) is the same as that given in (8.15) but with unadjusted replaced by adjusted treatment means.

Since balanced incomplete block designs are efficiency-balanced, all main effect and interaction contrasts are estimated with the same precision. However, in a factorial experiment each treatment contrast need not be of equal importance. Interest may be primarily concerned with main effects and two-factor interactions, especially in experiments with many factors. High-order interactions may be of little importance or may even be assumed to be negligible. A design which provides more precise estimates of those contrasts of interest may therefore be preferred to a balanced incomplete block design. In any case, balanced incomplete block designs exist for only a limited number of parameter combinations so that it will be necessary for this reason also to consider setting out factorial experiments in other incomplete block designs.

9.3 Group divisible designs

Group divisible designs provide another important class of designs whose basic contrasts correspond to the main effects and interactions of a factorial experiment. A group divisible design has $v = mn$ treatments divided into m groups of n treatments each such that all pairs of treatments belonging to the same group occur together in λ_1 blocks, while pairs of treatments from different groups occur together in λ_2 blocks. Treatments in the same group are said to be *first associates* and those from different groups are said to be *second associates*. The arrangement of treatments into groups is called the *association scheme* of the design. It will be assumed that $\lambda_1 \neq \lambda_2$, since if $\lambda_1 = \lambda_2$ the design will be a balanced incomplete block design. The parameters v, k, r, b, m, n, λ_1 and λ_2 of a group divisible design must satisfy, in addition to $v = mn$ and (1.1),

$$r(k - 1) = (n - 1)\lambda_1 + n(m - 1)\lambda_2 \qquad (9.7)$$

This is a generalization of (1.2) for balanced incomplete block designs. Now a given treatment will occur with each of its $(n - 1)$ first associates in λ_1 blocks and with each of its $n(m - 1)$ second associates in λ_2 blocks; hence (9.7) is established. A group divisible design cannot exist unless both (1.1) and (9.7) are satisfied, although, as was the case with balanced incomplete block designs, such conditions are not sufficient for the existence of a group divisible design.

Example 9.2
Consider the following group divisible design with $v = 6$, $r = 2$, $k = 4$, $b = 3$, $m = 3$, $n = 2$:

$$
\begin{array}{ccc}
0 & 0 & 2 \\
1 & 1 & 3 \\
2 & 4 & 4 \\
3 & 5 & 5 \\
\end{array}
$$

The association scheme is given in Table 9.1.

The design has $\lambda_1 = 2$ and $\lambda_2 = 1$. For instance, treatments 0 and 1 are first associates and occur together in the first two blocks, while treatments 0 and 2 are second associates and occur together only in the first block.

9.3.1 Construction of group divisible designs

Many group divisible designs can be derived from balanced incomplete block designs. Let the labels $1, 2, \ldots, m$ be set out in a balanced incomplete block design with parameters $v^* = m$, k^*, r^*,

Table 9.1. *Association scheme of a group divisible design with six treatments*

Group	Treatments	
1	0	1
2	2	3
3	4	5

b^* and λ^*. Suppose that throughout this design label i is replaced by all n treatments in the ith group of the association scheme $(i = 1, \ldots, m)$. The resulting design will be a group divisible design with $v = mn$, $k = nk^*$, $r = r^*$, $b = b^*$, $\lambda_1 = r$ and $\lambda_2 = \lambda^*$. The design given in Example 9.2 is a group divisible design based on the unreduced balanced incomplete block design (1,2), (1,3), (2,3). Label 1 is replaced by treatments 0, 1 of group 1, labels 2 and 3 are replaced by 2, 3 and 4, 5 respectively.

In a balanced incomplete block design with parameters v^*, k^*, r^*, b^* and $\lambda = 1$, if all the blocks containing one particular treatment are omitted then the resulting design is a group divisible design with parameters

$$v = v^* - 1, \ r = r^* - 1, \ k = k^*, \ b = b^* - r^*,$$
$$m = r^*, \ n = k^* - 1, \ \lambda_1 = 0, \ \lambda_2 = 1$$

Example 9.3
Omitting all blocks containing treatment 12 from the design with 13 treatments in Example 1.2 gives

0	3	6	0	1	2	0	1	2
1	4	7	3	4	5	4	5	3
2	5	8	6	7	8	8	6	7
9	9	9	10	10	10	11	11	11

The association scheme of this group divisible design is given in Table 9.2.

Other designs can now be constructed by omitting all treatments from one or more groups of treatments. For instance, omitting the three treatments 9, 10 and 11 in the above design gives a design for nine treatments in nine blocks of three.

Table 9.2. *Association scheme of a group divisible design with 12 treatments*

Group	Treatments		
1	0	5	7
2	1	3	8
3	2	4	6
4	9	10	11

Many group divisible designs can also be obtained as the duals of square lattice designs and resolvable balanced incomplete block designs, including balanced lattice designs.

Numerous other methods have been used to construct group divisible designs; see, for instance, John (1971), Raghavarao (1971), Clatworthy (1973), Freeman (1976b) and John and Turner (1977). Group divisible designs belong to the wider class of partially balanced incomplete block designs with p associate classes introduced by Bose and Nair (1939). They comprise an important class within the family with $p = 2$, and have been extensively catalogued by Clatworthy (1973).

9.3.2 Factorial experiments

Let the association scheme of a group divisible design be such that the ith group consists of the n treatments

$$(i-1)n, (i-1)n+1, \ldots, in-1$$

for $i = 1, \ldots, m$. Then the concurrence matrix $\mathbf{NN}'$ can be partitioned into m^2 $n \times n$ submatrices. The diagonal submatrices have all diagonal elements equal to r and all off-diagonal elements equal to λ_1, while all elements of the off-diagonal submatrices are equal to λ_2, i.e.

$$\mathbf{NN}' = \mathbf{I}_m \otimes \{(r-\lambda_1)\mathbf{I}_n + \lambda_1\mathbf{J}_n\} + (\mathbf{J}_m - \mathbf{I}_m) \otimes \lambda_2\mathbf{J}_m$$

The information matrix $\mathbf{A}$ is then

$$\mathbf{A} = (1/k)\mathbf{I}_m \otimes \{(rk-r+\lambda_1)\mathbf{I}_n - (\lambda_1-\lambda_2)\mathbf{J}_n\} - (\lambda_2/k)\mathbf{J}_{mn} \quad (9.8)$$

Suppose that a factorial experiment with two factors F and G at m and n levels respectively is set out in a group divisible design. Let the levels of factor F correspond to the different treatment groups, and the levels of factor G to the treatments within the groups. That is, treatment $in + j$ is the treatment combination with factor F at the ith level and factor G at the jth level $(i = 0, 1, \ldots, m-1; \ j = 0, 1, \ldots, n-1)$. For example, in the association scheme of Table 9.1 treatment number 4 corresponds to the treatment combination with factors F and G at levels 2 and 0 respectively.

From (8.7) the main effect contrasts for factor F are given by

$$\mathbf{C}_{10} = (\mathbf{I}_m - \mathbf{K}_m) \otimes \mathbf{K}_n$$

With $\mathbf{A}$ given by (9.8), it can then be shown using (9.7) that

$$\mathbf{A}\mathbf{C}_{10} = re_1\mathbf{C}_{10}$$

where $e_1 = v\lambda_2/rk$. In other words, the main effect contrasts for factor F are basic contrasts of the group divisible design, with canonical efficiency factor e_1. Similarly it can be shown that the main effect contrasts for factor G and the FG interaction contrasts are basic contrasts, all with the same canonical efficiency factor $e_2 = (rk - r + \lambda_1)/rk$. The Moore–Penrose generalized inverse of $\mathbf{A}$ is then given by

$$\mathbf{A}^+ = (re_1)^{-1}\mathbf{C}_{10} + (re_2)^{-1}(\mathbf{C}_{01} + \mathbf{C}_{11}) \tag{9.9}$$

It follows that all main effects and interactions are mutually orthogonal and are estimated independently of each other. $\mathbf{C}_x\boldsymbol{\tau}$ is estimated by

$$\mathbf{C}_x\hat{\boldsymbol{\tau}} = (re_x)^{-1}\mathbf{C}_x\mathbf{q} \tag{9.10}$$

with

$$\mathrm{V}(\mathbf{C}_x\hat{\boldsymbol{\tau}}) = (re_x)^{-1}\mathbf{C}_x\sigma^2 \tag{9.11}$$

where $e_x = e_1$ for the F main effect and $e_x = e_2$ for the G main effect and the FG interaction. Apart from the efficiency factor, (9.10) and (9.11) are the same as the corresponding expressions for a balanced incomplete block design given by (9.4) and (9.5). Also, the sum of squares due to testing $H_0 : \mathbf{C}_x\boldsymbol{\tau} = 0$ is

$$S(H_0) = (re_x)^{-1}(\mathbf{C}_x\mathbf{q})'(\mathbf{C}_x\mathbf{q}) \tag{9.12}$$

which again is the same form as (9.6). The analysis of a two-factor experiment set out in a group divisible design can, therefore, be seen to be relatively straightforward.

More generally, it is possible to split factors F and G into subfactors to provide a more flexible class of designs for factorial experiments. For example, a group divisible design for 24 treatments in four groups of six could be used for a $2^3 \times 3$ experiment, with the F factor split into two factors F_1 and F_2 each at two levels, and the G factor split into two factors G_1 and G_2 at two and three levels respectively. All main effects and interactions are independently estimated by (9.10), with $e_x = e_1$ for the main effects of factors F_1 and F_2 and the F_1F_2 interaction and $e_x = e_2$ for the main effects of factors G_1 and G_2 and the F_1G_1, F_1G_2, F_2G_1, F_2G_2, G_1G_2, $F_1F_2G_1$ and remaining high-order interactions.

Since interactions are estimated with the same precision as one or other of the main effects, choosing a group divisible design to provide good estimates of main effects will necessarily mean that equal importance will be attached to the high-order interactions. Hence, these designs, although simple to use and analyse, have the same drawbacks as balanced incomplete block designs when used in factorial experiments.

9.4 Factorial structure

If a factorial experiment is set out in a complete block, balanced incomplete block or group divisible design then any treatment contrast belonging to one factorial effect (main effect or interaction) can be estimated independently of every contrast belonging to any other effect, i.e. factorial effects are mutually orthogonal. One consequence is that the adjusted treatment sum of squares can be partitioned orthogonally into sums of squares for each of the factorial effects. Hence, any effect can be estimated and assessed independently of any other effect. A further feature of such designs is that their canonical efficiency factors are directly associated with the different main effects and interactions, thereby permitting optimality criteria for choosing appropriate designs to be readily obtained. Incomplete block designs which possess these properties will be said to have (orthogonal) *factorial structure*. The problem of identifying which designs, or class of designs, have factorial structure will be considered in this section. It will be assumed that all designs are connected.

The v treatment combinations of an n-factor experiment with factor F_i at m_i levels $(i = 1, 2, \ldots, n)$ will be assumed to be in lexographical order, i.e. where the levels of the last factor are changed first. Now a design will have factorial structure, with respect to this ordering, if and only if treatment contrasts belonging to different factorial effects are uncorrelated, i.e. the covariance matrix given in (9.2) is zero. The adjusted treatment sum of squares $\hat{\tau}'\mathbf{q}$ can then be partitioned orthogonally as follows

$$\hat{\tau}'\mathbf{q} = \sum_{x \neq 0} \mathrm{S}(x)$$

where

$$\mathrm{S}(x) = (\mathbf{C}_x\hat{\tau})'(\mathbf{C}_x\mathbf{\Omega}\mathbf{C}_x)^{-}\mathbf{C}_x\hat{\tau} \tag{9.13}$$

is the sum of squares due to testing $H_0 : \mathbf{C}_x\boldsymbol{\tau} = \mathbf{0}$.

Mukerjee (1979) has established conditions for factorial structure in terms of the information matrix $\mathbf{A}$ of the design. First, define a proper matrix to be a square matrix with all row sums and all column sums equal. Now let the information matrix of the factorial design be given by

$$\mathbf{A} = \sum_{i=1}^{w} \xi_i (\mathbf{V}_{i1} \otimes \mathbf{V}_{i2} \otimes \ldots \otimes \mathbf{V}_{in}) \tag{9.14}$$

where w is a positive integer, $\xi_1, \xi_2, \ldots, \xi_w$ are real numbers and, for each i, $\mathbf{V}_{ij}$ is a proper matrix of order m_j $(j = 1, 2, \ldots, n)$. Mukerjee (1979) shows that a sufficient condition for a design to have factorial structure is that its information matrix be of the form given by (9.14). The condition is also shown to be necessary if the design is connected.

Sufficiency can be established as follows. With $\mathbf{C}_x$ given by (8.10) and $\mathbf{A}$ by (9.14) it follows that

$$\mathbf{C}_x \mathbf{A} = \mathbf{A} \mathbf{C}_x \tag{9.15}$$

since $\mathbf{VK} = \mathbf{KV}$ for any proper matrix $\mathbf{V}$. Then, using (1.13),

$$\mathbf{C}_x \mathbf{A} \hat{\boldsymbol{\tau}} = \mathbf{A} \mathbf{C}_x \hat{\boldsymbol{\tau}} = \mathbf{C}_x \mathbf{q}$$

so that

$$\mathbf{C}_x \hat{\boldsymbol{\tau}} = \boldsymbol{\Omega} \mathbf{C}_x \mathbf{q}$$

where $\boldsymbol{\Omega}$ is any generalized inverse of $\mathbf{A}$. Hence, as $V(\mathbf{q}) = \mathbf{A}\sigma^2$ and $\mathbf{C}_x \mathbf{C}_y = \mathbf{0}$ $(x \neq y)$ it is seen that

$$\mathrm{cov}(\mathbf{C}_x \hat{\boldsymbol{\tau}}, \mathbf{C}_y \hat{\boldsymbol{\tau}}) = \boldsymbol{\Omega} \mathbf{C}_x \mathbf{A} \mathbf{C}_y \boldsymbol{\Omega}' \sigma^2 = \mathbf{0}$$

for $x \neq y$. Thus a design whose information matrix $\mathbf{A}$ is given by (9.14) has factorial structure.

It can also be established that the canonical efficiency factors of a design with factorial structure can be identified with the main effects and interactions of the factorial experiment. Let $\mathbf{A}$ be written in canonical form as $\mathbf{A} = \sum \theta_i \mathbf{p}_i \mathbf{p}_i'$ where $\mathbf{p}_i$ is a normalized eigenvector of $\mathbf{A}$ with non-zero eigenvalue θ_i. Then using (9.15),

$$\mathbf{C}_x \mathbf{A} = \sum \theta_i (\mathbf{C}_x \mathbf{p}_i)(\mathbf{C}_x \mathbf{p}_i')$$

so that $\mathbf{C}_x \mathbf{A}$ can be written as

$$\mathbf{C}_x \mathbf{A} = \sum_i \theta_{xi} \mathbf{t}_{xi} \mathbf{t}_{xi}' \tag{9.16}$$

where the $\mathbf{t}_{xi}$ are distinct vectors obtained from the $\mathbf{C}_x \mathbf{p}_i$ and normalized so that

$$\mathbf{t}_{xi}' \mathbf{t}_{xj} = \begin{cases} 1, & i = j \\ 0, & i \neq j \end{cases}$$

Again using (9.15),

$$\mathbf{C}_x \mathbf{A} \mathbf{C}_y \mathbf{A} = \sum_i \sum_j \theta_{xi}\theta_{yj} \mathbf{t}_{xi} \mathbf{t}'_{xi} \mathbf{t}_{yj} \mathbf{t}'_{yj} = 0 \quad (x \neq y)$$

which implies that $\mathbf{t}_{xi} \neq \mathbf{t}_{yj}$ for all i and j. Hence

$$\mathbf{A} = \sum_x \mathbf{C}_x \mathbf{A} = \sum_x \sum_i \theta_{xi} \mathbf{t}_{xi} \mathbf{t}'_{xi} \qquad (9.17)$$

where the $\mathbf{t}_{xi}$ are distinct vectors for all x and i. Hence, $\mathbf{t}_{xi}$ is an eigenvector of $\mathbf{A}$ and is in the contrast space spanned by $\mathbf{C}_x$. The linear function $\mathbf{t}'_{xi}\boldsymbol{\tau}$ is, therefore, a basic contrast of the design and represents one degree of freedom of the generalized interaction $\mathbf{C}_x\boldsymbol{\tau}$ with corresponding canonical efficiency factor $e_{xi} = \theta_{xi}/r$. The $v-1$ canonical efficiency factors are thus identified with the main effects and interactions of the factorial experiment. An average efficiency factor, E_x, can then be defined for the generalized interaction as

$$E_x = \nu_x / \sum e_{xi}^{-1} \qquad (9.18)$$

where ν_x is the number of degrees of freedom of the interaction.

The generalized interaction $\mathbf{C}_x\boldsymbol{\tau}$ is estimated by $\mathbf{C}_x\hat{\boldsymbol{\tau}} = \mathbf{C}_x\boldsymbol{\Omega}\mathbf{q}$ with $V(\mathbf{C}_x\hat{\boldsymbol{\tau}}) = \mathbf{C}_x\boldsymbol{\Omega}\mathbf{C}_x\sigma^2$. The sum of squares (9.13) can be easily calculated. Using (9.15) and the estimability condition $\mathbf{C}_x = \mathbf{C}_x\boldsymbol{\Omega}\mathbf{A}$ it can be shown that $\mathbf{A}$ is a generalized inverse of $\mathbf{C}_x\boldsymbol{\Omega}\mathbf{C}_x$. Therefore, again using (9.15),

$$S(x) = \hat{\boldsymbol{\tau}}'\mathbf{C}_x\mathbf{A}\mathbf{C}_x\hat{\boldsymbol{\tau}} = \hat{\boldsymbol{\tau}}'\mathbf{C}_x\mathbf{A}\hat{\boldsymbol{\tau}}$$

so that, since $\mathbf{C}_x$ is idempotent and $\mathbf{A}\hat{\boldsymbol{\tau}} = \mathbf{q}$,

$$S(x) = (\mathbf{C}_x\hat{\boldsymbol{\tau}})'(\mathbf{C}_x\mathbf{q}) \qquad (9.19)$$

Once the incomplete block analysis has been carried out, the further partitioning of the adjusted treatment sum of squares into components due to main effects and interactions is particularly straightforward. The formula given in (9.19) is similar in form to that given in (8.15) for complete block designs, in (9.6) for balanced incomplete block designs and in (9.12) for group divisible designs. Instead of calculations being based on sums of squares of unadjusted or adjusted treatment totals, (9.19) involves the sum of cross products of adjusted treatment totals and estimates of treatment effects.

Many different classes of designs have factorial structure. They include complete block, balanced incomplete block, and group divisible designs; all have information matrices which involve the Kronecker products of $\mathbf{I}$ and $\mathbf{J}$ matrices and, hence, satisfy the condition of (9.14). Also included are all designs satisfying property A of Kurkjian and Zelen (1963) and the n-cyclic designs of Section 3.5. Property A designs will be considered in the next section and n-cyclic designs in Section 9.6.

9.5 Balance

If a design has factorial structure then it has been shown that there exists $v - 1$ linearly independent basic contrasts which span the contrast spaces defined by the main effects and interactions of the factorial experiment. The basic contrast $\mathbf{t}'_{xi}\boldsymbol{\tau}$ is estimated by

$$\mathbf{t}'_{xi}\hat{\boldsymbol{\tau}} = (re_{xi})^{-1}\mathbf{t}'_{xi}\mathbf{q}$$

with

$$\mathrm{var}(\mathbf{t}'_{xi}\hat{\boldsymbol{\tau}}) = (re_{xi})^{-1}\sigma^2$$

Further, the sum of squares $\mathrm{S}(x)$ given in (9.13) can be partitioned orthogonally into single degree of freedom components. Thus

$$\mathrm{S}(x) = \sum_i (re_{xi})^{-1}(\mathbf{t}'_{xi}\mathbf{q})^2 \tag{9.20}$$

If the canonical efficiency factors e_{xi} corresponding to a particular generalized interaction are all distinct then this decomposition will be unique. At the other extreme, if all canonical efficiency factors are equal then any orthogonal decomposition is possible. A design is said to be *balanced* with respect to the generalized interaction $\mathbf{C}_x\boldsymbol{\tau}$ if and only if $e_{xi} = E_x$ for all i, where E_x is given in (9.18).

If a design is balanced with respect to all main effects and interactions it will be said to have *factorial balance*. Such designs have a particularly simple analysis and also permit any single degree of freedom orthogonal decomposition of main effect and interaction sums of squares. Clearly, any design which is efficiency-balanced also has factorial balance. It follows that complete block and balanced incomplete block designs have factorial balance. Group divisible designs, on the other hand, are not efficiency-balanced but do have factorial balance, as was shown in Section 9.3.2. Kshirsagar (1966) has proved that a design has factorial balance if and only if it satisfies property A of Kurkjian and Zelen (1963), i.e. is a design whose information matrix is of the form

$$\mathbf{A} = \sum_s h(s_1, s_2, \ldots, s_n)(\mathbf{D}_{s_1} \otimes \mathbf{D}_{s_2} \otimes \ldots \otimes \mathbf{D}_{s_n}) \tag{9.21}$$

where each $s_j = 0$ or 1,

$$\mathbf{D}_{s_j} = \left\{ \begin{array}{ll} \mathbf{K}_{m_j}, & s_j = 1 \\ \mathbf{I}_{m_j}, & s_j = 0 \end{array} \right.$$

where the $h(s_1, s_2, \ldots, s_n)$ are constants depending on the s_j, and where the summation is over all binary numbers $s = (s_1 s_2 \ldots s_n)$. Such designs clearly have factorial structure since the information matrix has the form given by (9.14). It can be verified that the columns of $\mathbf{C}_x$ are eigenvectors of $\mathbf{A}$, so that the canonical efficiency factors are given by

$$E_x = (1/r) \sum h(s_1, s_2, \ldots, s_n) \tag{9.22}$$

where the summation now is over the subset of binary numbers $s = (s_1 s_2 \ldots s_n)$ defined by fixing $s_i = 0$ if $x_i = 1$ $(i = 1, 2, \ldots, n)$. For example, in a two-factor experiment $rE_{10} = h(0,0) + h(0,1)$ and $rE_{11} = h(0,0)$.

Included in this class are complete block, balanced incomplete block and group divisible designs. The form of the $\mathbf{A}$ matrix in (9.21) can be used to check whether a design has factorial balance but it does not provide a general method of constructing such designs. Even when a factorial balanced design is given by a balanced incomplete block or group divisible design it may be relatively inefficient in estimating the important factorial effects. The class of *n*-cyclic designs provides a simple method of constructing a large number of designs suitable for factorial experiments. Many of these designs have desirable properties with regard to both efficiency factors and balance. These designs will now be discussed.

9.6 *n*-Cyclic designs

In an *n*-cyclic design a treatment is represented by an *n*-tuple $a = a_1 a_2 \ldots a_n$ where $a_i = 0, 1, \ldots, m_i - 1$ $(i = 1, 2, \ldots, n)$. Designs are obtained by cyclical development of one or more initial blocks as described in Section 3.5. The information matrix of an *n*-cyclic design is, following (3.8),

$$\mathbf{A} = \sum_{h_1=0}^{m_1-1} \sum_{h_2=0}^{m_2-1} \cdots \sum_{h_n=0}^{m_n-1} \xi_{h_1 h_2 \ldots h_n} (\mathbf{\Gamma}_{h_1} \otimes \mathbf{\Gamma}_{h_2} \otimes \ldots \otimes \mathbf{\Gamma}_{h_n}) \tag{9.23}$$

where $\xi_{h_1 h_2 \ldots h_n}$ are elements of the first row of $\mathbf{A}$ and where $\mathbf{\Gamma}_{h_i}$ is an $m_i \times m_i$ circulant matrix whose first row has 1 in the $(h_i + 1)$th column and zero elsewhere. Note that, since the eigenvectors of a circulant matrix do not depend on the individual elements of that matrix, there will exist a set of eigenvectors of the matrix $\mathbf{A}$ in (9.23) of the form

$$\boldsymbol{\gamma}_1 \otimes \boldsymbol{\gamma}_2 \otimes \ldots \otimes \boldsymbol{\gamma}_n \qquad (9.24)$$

where $\boldsymbol{\gamma}_i$ is a vector of length m_i; see Section A.8 of the Appendix.

Now let the treatment $a = a_1 a_2 \ldots a_n$ correspond to a treatment combination in an n-factor experiment where a_i represents a level of factor F_i $(i = 1, 2, \ldots, n)$. It then follows that all n-cyclic designs have factorial structure, since the $\mathbf{A}$ matrix given in (9.23) has the same form as that given in (9.14). n-Cyclic designs, therefore, provide a flexible class of incomplete block designs suitable for use in factorial experiments.

9.6.1 Canonical efficiency factors

For a factorial experiment of given size, a number of alternative n-cyclic designs will usually exist. The choice of an appropriate design can again be based on the canonical efficiency factors, and also on their degree of balance. The canonical efficiency factors are, from (3.9), given by

$$e_{u_1 \ldots u_n} = (1/r) \sum_{h_1} \cdots \sum_{h_n} \xi_{h_1 \ldots h_n} \cos \left\{ \sum_{j=1}^{n} (2\pi u_j h_j / m_j) \right\} \qquad (9.25)$$

$$(u_l = 0, 1, \ldots, m_l - 1; \; l = 1, 2, \ldots, n)$$

Excluded from (9.25) is the case where $u_l = 0$ for all l since $e_{00 \ldots 0} = 0$ is the eigenvalue of $\mathbf{A}$ corresponding to the vector $\mathbf{1}$.

With $u_l = 0$ for $l = 2, 3, \ldots, n$ and letting u_1 take non-zero values, (9.25) provides $m_1 - 1$ canonical efficiency factors given by

$$e_{u_1 0 \ldots 0} = (1/r) \sum_{h_1} \xi_{h_1} \cos(2\pi u_1 h_1 / m_1) \qquad (9.26)$$

where

$$\xi_{h_1} = \sum_{h_2} \cdots \sum_{h_n} \xi_{h_1 h_2 \ldots h_n}$$

It can be seen, from a comparison with (3.2), that (9.26) also provides the canonical efficiency factors of a cyclic design whose information matrix is a circulant matrix with first-row elements given by ξ_{h_1} $(h_1 = 0, 1, \ldots, m_1 - 1)$. Further, if $\boldsymbol{\gamma}_1$ is an eigenvector of the information matrix of this cyclic design, then $\boldsymbol{\gamma}_1 \otimes \mathbf{1} \otimes \ldots \otimes \mathbf{1}$ is an eigenvector of the $\mathbf{A}$ matrix given in (9.23). Hence, the canonical efficiency factors given by (9.26) are those associated with the main effect of factor F_1.

If $u_l = 0$ for $l = 3, 4, \ldots, n$ and if u_1 and u_2 take zero and non-zero values, other than $u_1 = u_2 = 0$, then (9.25) provides $m_1 m_2 - 1$ canonical efficiency factors given by

$$e_{u_1 u_2 0 \ldots 0} = (1/r) \sum_{h_1} \sum_{h_2} \xi_{h_1 h_2} \cos\{(2\pi u_1 h_1 / m_1) + (2\pi u_2 h_2 / m_2)\}$$

(9.27)

where

$$\xi_{h_1 h_2} = \sum_{h_3} \cdots \sum_{h_n} \xi_{h_1 h_2 \ldots h_n}$$

Again (9.27) provides the canonical efficiency factors of a 2-cyclic design whose information matrix has elements $\xi_{h_1 h_2}$. If $\boldsymbol{\gamma}_1 \otimes \boldsymbol{\gamma}_2$ is an eigenvector of this 2-cyclic design then $\boldsymbol{\gamma}_1 \otimes \boldsymbol{\gamma}_2 \otimes \mathbf{1} \otimes \ldots \otimes \mathbf{1}$ is an eigenvector of the $\mathbf{A}$ matrix given in (9.23). Hence, the canonical efficiency factors of (9.27) correspond to the F_1 main effect when $u_2 = 0$, the F_2 main effect when $u_1 = 0$, and to the $F_1 F_2$ interaction when both u_1 and u_2 are non-zero.

More generally, the canonical efficiency factors associated with the main effects and interactions of factors $F_1, F_2, \ldots, F_s$ ($s < n$) are obtained from (9.25) by putting $u_l = 0$ for $l = s+1, \ldots, n$ and letting the other values be zero and non-zero. These factors will also be the canonical efficiency factors of an s-cyclic design whose information matrix has elements

$$\xi_{h_1 h_2 \ldots h_s} = \sum_{h_{s+1}} \cdots \sum_{h_n} \xi_{h_1 h_2 \ldots h_n}$$

(9.28)

The s-cyclic design whose information matrix has elements given by (9.28) can be obtained from the full n-cyclic design by first deleting from each treatment combination the last $n - s$ factors and then deleting any duplicate blocks from the resulting design. To show this let $\mathbf{N}$, $\mathbf{N}_s$ and $\mathbf{A}$, $\mathbf{A}_s$ be the incidence and information matrices of the n-cyclic and derived s-cyclic designs respectively. Now $\mathbf{N}_s$ is obtained from $\mathbf{N}$ by summing the rows of $\mathbf{N}$ over the last $n - s$ factors so that the relationship between $\mathbf{N}$ and $\mathbf{N}_s$ can be represented by

$$(\mathbf{I} \otimes \mathbf{K})\mathbf{N} = \mathbf{N}_s \otimes \mathbf{K}$$

(9.29)

where $\mathbf{I}$ and $\mathbf{K}$ are of order $m_1 m_2 \ldots m_s$ and $m_{s+1} \ldots m_n$ respectively. The presence of the $\mathbf{K}$ matrix on the right-hand side of (9.29) is due to duplicate blocks existing when factors are deleted from the n-cyclic design; these are eliminated in the s-cyclic design. It follows from (9.29) that

$$(\mathbf{I} \otimes \mathbf{K})\mathbf{A}(\mathbf{I} \otimes \mathbf{K}) = (\mathbf{A}_s \otimes \mathbf{K})$$

and since, with $\mathbf{A}$ given by (9.23),

$$(\mathbf{I} \otimes \mathbf{K})\mathbf{A}(\mathbf{I} \otimes \mathbf{K}) = \left\{ \sum_{h_1} \cdots \sum_{h_s} \xi_{h_1 \ldots h_s} (\mathbf{\Gamma}_{h_1} \otimes \ldots \otimes \mathbf{\Gamma}_{h_s}) \right\} \otimes \mathbf{K}$$

the result is established.

Hence, the canonical efficiency factors of the n-cyclic design corresponding to the main effects and interactions of factors $F_1, F_2, \ldots, F_s$ can be obtained either from (9.25) by putting $u_l = 0$ for $l = s+1, \ldots, n$, or as the canonical efficiency factors of the s-cyclic design with incidence matrix $\mathbf{N}_s$. The above results are set out in terms of the first s factors so as to avoid making the notation unnecessarily complicated, but these results apply equally well to any set of s factors.

Example 9.4
Consider the 3-cyclic design for a $4 \times 3 \times 2$ factorial experiment given by the initial block (000 001 221 310). Deleting, for instance, the second and third factors and eliminating duplicate blocks gives a cyclic design with four treatments having initial block (0 0 2 3). The canonical efficiency factors of this cyclic design are equal to those of the F_1 main effect in the three-factor design. Deleting only the second factor leads to a 4×2, 2-cyclic design with initial block (00 01 21 30). Its canonical efficiency factors are those of the F_1 and F_3 main effects and $F_1 F_3$ interaction in the three-factor experiment.

In addition to the canonical efficiency factors, the degree of balance in an n-cyclic design can also be determined from the derived s-cyclic design. Since eigenvalues of $\mathbf{A}_s$ are also eigenvalues of $\mathbf{A}$, the n-cyclic design will be balanced with respect to all main effects and interactions involving the s factors if and only if the derived s-cyclic design has factorial balance.

Lewis and Dean (1985) have shown that this correspondence between the efficiency factors of the main effects and interactions of a subset s of the n factors and the efficiency factors of a derived s-factor design holds for all equally replicated designs with factorial structure.

9.6.2 *Construction of n-cyclic factorial designs*

It has been shown how the canonical efficiency factors of main effects, two-factor, three-factor,... interactions in an n-cyclic design can be directly obtained from cyclic, 2-cyclic, 3-cyclic,... designs. However, instead of breaking down the n-cyclic designs to obtain the canonical efficiency factors the reverse process can be used to build up the n-factor design. Cyclic designs can be merged to provide 2-cyclic designs, these in turn can be merged to give 3-cyclic designs, and so on.

Consider the main effect of factor F_1. Its $m_1 - 1$ canonical efficiency factors can be obtained directly from (9.26) or from the cyclic design whose initial blocks are obtained from the initial blocks of the n-cyclic design by deleting all factors except the first. All $m_1 - 1$ canonical efficiency factors are equal if the cyclic design is balanced. Further, its average efficiency factor $E(10...0)$ is equal to the average efficiency factor E_1 of the cyclic design. The same results apply to any main effect. Hence optimal main effect n-cyclic designs are built up from optimal cyclic designs.

For the $F_1 F_2$ interaction attention is focused on the 2-cyclic design obtained from the n-cyclic design by deleting all factors except the first two. This 2-cyclic design has $m_1 m_2 - 1$ canonical efficiency factors with $m_1 - 1$ and $m_2 - 1$ of them associated with the main effects of F_1 and F_2 respectively and the remaining $(m_1 - 1)(m_2 - 1)$ factors associated with the $F_1 F_2$ interaction. If they are all equal then the n-cyclic design is balanced with respect to this interaction. It follows that the cyclic designs chosen for the main effects of factors F_1 and F_2 should be merged so as to produce the most efficient 2-cyclic design if maximum precision is required for the $F_1 F_2$ interaction. The following three examples illustrate the procedure.

Example 9.5

Consider the construction of a design for a 4^2 experiment in 16 blocks of six units per block. For both of the main effects, a cyclic design for four treatments in blocks of six is required. Two possible designs are given by initial blocks (0 0 0 1 2 3) and (0 0 1 1 2 3). The first design is balanced with average efficiency factor 0.89, the second design is not balanced but is A-optimal with average efficiency factor 0.96. A number of 2-cyclic designs can be obtained by merging these cyclic designs. One such design is given by the initial block (00 00 11 11 22 33); note that deleting either factor produces a cyclic design with initial block (0 0 1 1 2 3). Hence, this 2-cyclic design is obtained by using the second cyclic design above for both factors. Even though the average efficiency factors for both main effects are necessarily 0.96, the average efficiency factor for the interaction will be low as this 2-cyclic design is relatively inefficient. A better 2-cyclic design based on the same cyclic designs is given by initial block (00 01 10 12 23 31); the average interaction efficiency factor being $E(11) = 0.831$. Although this design does not have factorial balance it is worth noting that in this case each factor could be represented by two pseudo factors each at two levels and the design analysed as a factorial balanced 2^4 experiment.

As an alternative to the above, the 2-cyclic design with initial block (00 01 02 10 23 30) is a balanced incomplete block design and has been listed as such in Fisher and Yates (1963); there it is called a dicyclic design. The full design is, in fact, given in Example 9.1. It is based on the first of the two cyclic designs above and so all efficiency factors are 0.89. The other 2-cyclic design with initial block (00 01 10 12 23 31) may, however, be preferred if greater precision on main effects is required, at the expense of some loss in precision on the interaction. Such considerations become even more important as the number of factors increases, since the high-order interactions are rarely of much practical importance.

Example 9.6
Suppose a design is required for a 3×2^2 experiment in 12 blocks of four units per block with main effects and two-factor interactions estimated with as high a degree of precision as possible. Firstly, for the main effects efficient cyclic designs for three and two treatments respectively in blocks of four are required. For three treatments the most efficient cyclic design has initial block (0 0 1 2); it is balanced with efficiency factor 0.94. For two treatments a fully efficient cyclic design is given by the initial block (0 0 1 1).

The next step is to merge these cyclic designs to produce 2-cyclic designs. For the $F_1 F_2$ interaction a possible 3×2, 2-cyclic design, merging the cyclic designs (0 0 1 2) and (0 0 1 1), is given by the initial block (00 01 10 21); it has factorial balance. Another possible 2-cyclic design is given by the initial block (00 01 11 20); it is identical to the first one but will be used in the final 3-cyclic design for the $F_1 F_3$ interaction. For the $F_2 F_3$ interaction the 2^2, 2-cyclic design with initial block (00 01 10 11) is a complete block design and, hence, is fully efficient.

The final 3-cyclic design is obtained by merging the above cyclic

Table 9.3. *Efficiency factors for* 3×2^2 *designs*

Generalized interaction	3-cyclic design	Group divisible design
F_1	0.94	0.75
F_2	1.00	0.75
F_3	1.00	0.87
$F_1 F_2$	0.81	0.75
$F_1 F_3$	0.81	0.87
$F_2 F_3$	1.00	0.87
$F_1 F_2 F_3$	0.44	0.87
Overall	0.75	0.81

and 2-cyclic designs. Any design obtained in this way will have high efficiency factors, and will be balanced, with respect to all main effects and two-factor interactions. Different choices of such 3-cyclic designs will only affect properties of the $F_1 F_2 F_3$ interaction. One possible design is obtained from the initial block (000 011 101 210) and can be shown to have factorial balance.

An alternative factorial balanced design is provided by a group divisible design with $m = 6$ and $n = 2$; it is design R109 in the catalogue by Clatworthy (1973). The average efficiency factors for this group divisible design and for the above 3-cyclic designs are given in Table 9.3. Although the group divisible design has the higher overall efficiency factor, the 3-cyclic design may be preferred in a 3×2^2 factorial experiment since it is more efficient in estimating the main effects and two-factor interactions.

Example 9.7
n-Cyclic designs may be constructed when alternative balanced incomplete block and group divisible designs are not available; the reverse will also be true. One such example is a 5×4 experiment in 20 blocks of four. The 2-cyclic design with initial block (00 11 23 32) has factorial balance and is based on the A-optimal cyclic designs for both five and four treatments in blocks of four.

Efficient binary cyclic designs can be obtained from the tables or computer algorithms referred to in Section 3.4, and these can be used in the first stages of the construction of n-cyclic designs aimed at maximizing main effect efficiency factors. As has been seen in the above examples, the cyclic designs needed for the n-cyclic designs are often non-binary. In such cases it is suggested that the binary designs are augmented by complete blocks. Certainly such designs are (M,S)-optimal. It has been conjectured by John and Williams (1982) that an A-optimal binary design augmented by complete blocks will produce A-optimal non-binary designs. In Example 9.5 the A-optimal binary cyclic design (0 1) for $k = 2$ is augmented by a complete block design to give the A-optimal non-binary cyclic design (0 1 0 1 2 3) for $k = 6$.

Bearing in mind the requirement for (M,S)-optimality, a general rule in constructing cyclic, 2-cyclic, ... designs would be to ensure that treatment combinations occur as equally as possible in the initial block. In Example 9.5 the 2-cyclic design with initial block (00 00 11 11 22 33) is relatively inefficient because some treatment combinations occur twice, some once and others not at all in the initial block. A binary 2-cyclic design would be better in this case.

The designs in all three examples above have $b = v$ blocks. n-Cyclic designs with fewer blocks can be constructed if use is made of partial sets. The method of constructing such sets is given in Section 3.5. Basically the initial block will consist of a subgroup of the v treatment combinations of size d together with a number of its cosets, where d is a common divisor of v and k. The use of partial sets will, in general, impose further constraints on the efficiency factors of the main effects and interactions. For example, a 2-cyclic design for a 4^2 experiment in eight blocks of six is given by the initial block (00 20 01 21 12 32). Compared with the design in 16 blocks, there is a decrease in the average efficiency factors on both main effects, with the decrease being greatest on the F_2 effect since the cyclic set (0 0 1 1 2 2) has had to be used.

Finally, a computer program can be easily written which will enable the properties of competing n-cyclic designs to be readily compared. The main steps of such a program are as follows:

1. Input the parameters and initial blocks of the n-cyclic design.

2. Obtain the first rows of the concurrence matrix $\mathbf{NN}'$ and the information matrix $\mathbf{A}$ from the initial blocks using a method of differencing similar to that described in Section 3.4 for cyclic designs.

3. Calculate the canonical efficiency factors, and average efficiency factors, of each main effect and interaction using (9.25).

9.6.3 *Example of a two-factor experiment*

The analysis of a factorial experiment involving two factors at three and five levels respectively in 15 blocks of five will now be given. The design used was a binary 2-cyclic design generated from the initial block (00 01 12 14 23). It can be shown to have factorial balance with efficiency factors

$$E(10) = 0.96, \qquad E(01) = 1.00, \qquad E(11) = 0.76$$

Although factorial balance enables considerable simplifications to be made in the analysis, the method of analysis given below is one that is appropriate for any design with factorial structure. The starting point for the analysis of the factorial experiment is the intra-block treatment estimates $\hat{\tau}$ obtained by solving the reduced normal equations $\mathbf{A}\hat{\tau} = \mathbf{q}$ given in (1.13).

The experiment could, of course, be analysed using analysis of variance or regression facilities available on many statistical computer packages; for instance, the experiment presented here is readily analysed using the ANOVA directive in the GENSTAT package.

The experiment was a pilot study for a larger study concerned with modelling annoyance due to combinations of noises; the results of the larger study are given in Rice and Izumi (1984). The noises were made up of three *sounds* each at five *intensity* levels. The sounds consisted of a single steady continuous sound, namely road traffic (R), and two intermittent discrete sounds, trains (T) and aircraft (A). The main purpose of the pilot study was to see if there were differences in annoyance between reactions to the continuous sound and the discrete sounds, with the aim of eliminating one of the discrete sounds from the larger study.

The second factor was the intensity of the sounds expressed in terms of the equivalent A-weighted sound pressure levels (L_{Aeq}). Five equally spaced levels were used, namely 35, 44, 53, 62 and 71 L_{Aeq}.

The experiment was carried out in a simulated domestic living room listening facility at the Institute of Sound and Vibration Research, University of Southampton. The noises were played through multiple loudspeakers concealed behind the walls (for traffic and trains) and ceiling (for aircraft).

The 15 noises were presented to 15 subjects according to the design of Table 9.4; where, for example, A35 represents the aircraft sound at 35 L_{Aeq}. The order within columns gives the order in

Table 9.4. *Design for* 3 × 5 *experiment (after randomization)*

							Subject							
1	2	3	4	5	6	7	8	9	10	11	12	13	14	15
A62	R62	A44	R53	T35	T62	R35	T71	A53	R71	T53	A35	A71	T44	R44
T44	A44	T53	A35	R53	R71	A71	R35	T35	A62	R44	T71	T62	R62	A53
R71	T35	R62	T62	A71	A44	T44	A62	R44	T53	A35	R53	R35	A53	T71
A53	R53	A35	R71	T44	T35	R62	T53	A71	R44	T62	A62	A44	T71	R35
R35	T71	R71	T44	A62	A53	T53	A44	R62	T35	A71	R44	R53	A35	T62

which the noises were presented to the subjects. The duration of each noise was 10 minutes. During this time the road traffic sounds were heard continuously and the discrete sounds were heard at random intervals on five occasions, each discrete sound lasting about 30 seconds.

There was a short break between the presentation of each noise, during which the subjects reported their judged annoyance on a 0–9 numerical category scale, with the endpoints labelled "not at all annoying" and "extremely annoying"; these *subjective scale values* are the responses from the experiment.

The 2-cyclic design was set out in a row by column array with blocks corresponding to the columns of the array and with each row containing a single replicate of each treatment combination. The different levels of the sound and intensity factors were randomly allocated to the treatment combination labels, and subjects were randomly assigned to the columns of the array. The order of presentation of the treatment combinations to subjects was taken as a second blocking factor, and was randomly assigned to the rows of the array. The full design after randomization is given in Table 9.4. The subjective scale values together with subject and treatment combination totals are set out in Table 9.5.

The adjusted treatment totals **q** and the intra-block treatment estimates $\hat{\tau}$ obtained from the analysis of the block design with 15 treatments in 15 blocks of five are set out in Tables 9.6 and 9.7 in the form of two-way tables classified according to the treatment factors; margin totals are also given. Note that margin intensity totals in Table 9.6 are $r = 5$ times those in Table 9.7; this being a consequence of the main effect of intensity being estimated with full efficiency, i.e. $E(01) = 1.00$. The intra-block analysis of variance

Table 9.5. *Subjective scale values*

Noise	1	2	3	4	5	6	7	8	9	10	11	12	13	14	15	Totals
A35			2	2							3	1		0		8
A44		2	3		4		3					4				16
A53	5				5			3					3	3		19
A62	5			6			5		5		3					24
A71				8		8		8		7		6				37
R35	1					2	2					4		1		10
R44								1	7	7	2			2		19
R53		6		6	7						6	5				30
R62		7	8			9		6					9			39
R71	9		9	8	9				9							44
T35		0		2	2			1	1							6
T44	4			0	2		4						2			12
T53			6				4	6		2	4					22
T62				7		7					5		7		5	31
T71		9						8				6		8	7	38
Totals	24	24	28	23	25	27	27	24	19	24	26	18	26	22	18	355

Table 9.6. *Adjusted treatment totals*

Sound	Intensity (L_{Aeq})					Total
	35	44	53	62	71	
A	−15.4	−9.8	−3.0	1.0	12.4	−14.8
R	−13.8	−2.0	6.8	15.0	18.8	24.8
T	−17.8	−12.2	−3.8	7.0	16.8	−10.0
Total	−47.0	−24.0	0.0	23.0	48.0	0.0

is given in Table 9.8, where a sum of squares due to the order of
presentation can be calculated directly from the order totals, since
this second blocking factor is orthogonal to treatments.

The adjusted treatment sum of squares can now be partitioned
into orthogonal components representing the two main effects and
the interaction. The calculations are, from (9.19), based on sums
of cross products of the various totals in Tables 9.6 and 9.7. They
are

Table 9.7. *Intra-block treatment estimates*

Sound	Intensity (L_{Aeq})					Total
	35	44	53	62	71	
A	−2.90	−1.91	−0.63	−0.06	2.41	−3.08
R	−2.91	−0.29	1.52	3.19	3.66	5.17
T	−3.59	−2.60	−0.89	1.47	3.52	−2.08
Total	−9.40	−4.80	0.00	4.60	9.60	0.00

Table 9.8. *Intra-block analysis of variance*

Source of variation	d.f.	s.s.
Treatments (adjusted)	14	431.337
Subjects (unadjusted)	14	28.667
Order of presentation	4	10.267
Residual	42	58.396
Total	74	528.667

S(Sound)
$$= \tfrac{1}{5}\{(-14.8 \times -3.08) + (24.8 \times 5.17) + (-10.0 \times -2.08)\}$$
$$= 38.920$$

S(Intensity)
$$= \tfrac{1}{3}\{(-47 \times -9.4) + (-24 \times -4.8) + ... + (48 \times 9.6)\}$$
$$= 374.533$$

S(Sound by Intensity) $= 431.337 - 38.920 - 374.533 = 17.884$

Since intensity is a quantitative factor its main effect sum of squares can be further partitioned into linear, quadratic,... components. As the levels are equally spaced and the main effect is balanced these components can be estimated and tested using tables of orthogonal polynomials (see Fisher and Yates, 1963; or John and Quenouille, 1977, p. 36). More generally, the components are estimated using regression techniques. For five equally spaced levels the vectors of linear and quadratic coefficients are $\mathbf{L}' = (-2 \ -1 \ 0 \ 1 \ 2)$ and $\mathbf{Q}' = (2 \ -1 \ -2 \ -1 \ 2)$ respectively. Note that $\mathbf{L'L} = 10$ and $\mathbf{Q'Q} = 14$. The two sums of squares, linear intensity

Table 9.9. *Basic calculations for linear and quadratic components*

Sound	$\mathbf{L'q}$	$\mathbf{L'\hat{\tau}}$	$\mathbf{Q'q}$	$\mathbf{Q'\hat{\tau}}$
A	66.4	12.48	8.8	2.25
R	82.2	16.64	−16.6	−4.43
T	88.4	18.27	10.8	2.78
Total	237.0	47.40	3.0	0.60

(lin. int.) and quadratic intensity (quad. int.) are obtained from the quantities $\mathbf{L'q}$, $\mathbf{L'\hat{\tau}}$, $\mathbf{Q'q}$ and $\mathbf{Q'\hat{\tau}}$ calculated from the column totals in Tables 9.6 and 9.7; these quantities are given as the totals in Table 9.9. Hence,

$$S(\text{lin. int.}) \;=\; \tfrac{1}{10\times 3}(237.0 \times 47.4) = 374.46$$
$$S(\text{quad. int.}) \;=\; \tfrac{1}{14\times 3}(3.0 \times 0.6) = 0.043$$

In view of the structure of the sound factor its main effect sum of squares can be partitioned into two orthogonal components representing the comparison between the continuous and discrete sounds and a comparison among the discrete sounds respectively. The appropriate vectors of coefficients are $\mathbf{C'} = (-1\ 2\ -1)$ and $\mathbf{D'} = (1\ 0\ -1)$, with $\mathbf{C'C} = 6$ and $\mathbf{D'D} = 2$. The continuous against discrete sound (CD sound) and the within discrete sound (AT sound) sums of squares are obtained from the quantities $\mathbf{C'q}$, $\mathbf{C'\hat{\tau}}$, $\mathbf{D'q}$ and $\mathbf{D'\hat{\tau}}$ calculated from the row totals in Tables 9.6 and 9.7. Hence

$S(\text{CD sound})$
$$= \tfrac{1}{6\times 5}[\{(2 \times 24.8) - (-14.8) - (-10.0)\}$$
$$\times\{(2 \times 5.17) - (-3.08) - (-2.08)\}] = 38.44$$

$S(\text{AT sound})$
$$= \tfrac{1}{2\times 5}[\{-14.8 - (-10.0)\} \times \{-3.08 - (-2.08)\}] = 0.48$$

Since the design has factorial balance, the interaction sum of squares can also be partitioned into sums of squares which represent interactions of the sound components with the intensity components. These sum of squares are obtained from the basic calculations given in Table 9.9. They are

Table 9.10. *Full analysis of variance*

Source of variation	d.f.	s.s.	m.s.	F
Intensity	4	374.53	93.63	67.3
lin. int.	1	374.46	374.46	269.3
quad. int.	1	0.04	0.04	0.0
remainder	2	0.03	0.02	0.0
Sound	2	38.92	19.46	14.0
CD Sound	1	38.44	38.44	27.7
AT Sound	1	0.48	0.48	0.4
Intensity by Sound	8	17.88	2.24	1.6
lin. int. by CD Sound	1	0.41	0.41	0.3
quad. int. by CD Sound	1	8.73	8.73	6.3
lin. int. by AT Sound	1	6.37	6.37	4.6
quad. int. by AT Sound	1	0.04	0.04	0.0
remainder	4	2.34	0.59	0.4
Treatments (adjusted)	14	431.337	30.81	
Subjects (unadjusted)	14	28.667	2.05	
Order of presentation	4	10.267	2.57	
Residual	42	58.396	1.39	
Total	74	528.667		

S(lin. int. by CD sound)
$$= \tfrac{1}{6 \times 10}[\{(2 \times 82.2) - 66.4 - 88.4\} \times \{(2 \times 16.64) - 12.48 - 18.27\}] = 0.41$$

S(quad. int. by CD sound)
$$= \tfrac{1}{6 \times 14}[\{(2 \times -16.6) - 8.8 - 10.8\} \times \{(2 \times -4.43) - 2.25 - 2.78\}] = 8.73$$

S(lin. int. by AT sound)
$$= \tfrac{1}{2 \times 10}\{(66.4 - 88.4) \times (12.48 - 18.27)\} = 6.37$$

S(quad. int. by AT sound)
$$= \tfrac{1}{2 \times 14}\{(8.8 - 10.8) \times (2.25 - 2.78)\} = 0.04$$

The full analysis of variance is given in Table 9.10.

A table of adjusted means can be obtained, following (1.20), by adding the overall mean of the responses, namely 4.73, to the estimates in Table 9.7; these means are given in Table 9.11. The standard errors of the differences (SED) between adjusted means can be obtained from the Ω matrix of the incomplete

Table 9.11. *Adjusted means*

Sound	\multicolumn{5}{c}{Intensity (L_{Aeq})}	Total
	35	44	53	62	71	
A	1.83	2.82	4.11	4.67	7.15	4.12
R	1.82	4.44	6.25	7.92	8.40	5.77
T	1.15	2.14	3.84	6.20	8.25	4.32
Mean	1.60	3.13	4.73	6.27	7.93	4.73

block analysis. Mean standard errors can be obtained from the average efficiency factors; these standard errors will be exact for differences between main effect means since the design has factorial balance. The standard errors for means based on m observations are obtained from

$$\text{SED} = \sqrt{\left(\frac{2s^2}{mE}\right)}$$

where s^2 is the residual mean square and E the appropriate efficiency factor. For differences between sound and intensity adjusted means the standard errors are

$$\text{SED(Sound)} = \sqrt{\left(\frac{2 \times 1.39}{25 \times 0.96}\right)} = \pm 0.340$$

$$\text{SED(Intensity)} = \sqrt{\left(\frac{2 \times 1.39}{15 \times 1.0}\right)} = \pm 0.431$$

For differences between means in the main body of Table 9.11 the appropriate average efficiency factor is

$$E = \frac{14}{(2 \times 0.96^{-1}) + (4 \times 1) + (8 \times 0.76^{-1})} = 0.843$$

so that the average standard error is

$$\text{SED} = \sqrt{\left(\frac{2 \times 1.39}{5 \times 0.843}\right)} = \pm 0.812$$

The exact values are 0.821, 0.837 and 0.802 for differences between means from the same row, same column and different rows and columns respectively.

The conclusions from the analysis are that there are significant differences between intensity levels and between sounds. Further, differences between intensity levels are almost completely accounted for by the linear component. The estimate of the increase in subjective scale values per $9L_{Aeq}$ increase in intensity is

$$b = 47.4/(10 \times 3) = 1.58$$

There is also a significant difference between the continuous sound (road traffic) and the discrete sounds (aircraft and trains), but no difference between the two discrete sounds. Two components of the interaction are significant. From Table 9.11, it is clear that one component is due to the fact that the largest differences between the discrete and continuous sounds are in the middle intensity levels, while the other component is an indication of the fact that the average scale values for trains, although starting lower, increase more rapidly with intensity than aircraft. The final conclusion from the pilot study was that there was little to choose between the two discrete sounds; in the event, aircraft was chosen for the larger study.

It may be appropriate in some experiments to recover the inter-block information available on treatment comparisons. The method of obtaining combined intra- and inter-block treatment estimates is described in Chapter 7. If a computer program is available which provides these estimates then the calculation of adjusted treatment means and standard errors for the factorial experiment follows in a similar way to that given above. In the experiment presented here, however, the recovery of inter-subject information is unlikely to be worthwhile, as there is little or no between-subject information available on the important main effects.

9.7 Other designs

If $\mathbf{N}_i$ is the incidence matrix of a block design for m_i treatments in blocks of size k_i with each treatment replicated r_i times ($i = 1, 2, \ldots, n$), then

$$\mathbf{N} = \mathbf{N}_1 \otimes \mathbf{N}_2 \otimes \ldots \otimes \mathbf{N}_n \qquad (9.30)$$

is the incidence matrix of an n-factor design for $v = m_1 m_2 \ldots m_n$ treatment combinations in blocks of size $k = k_1 k_2 \ldots k_n$ with each treatment combination replicated $r = r_1 r_2 \ldots r_n$ times. The designs have factorial structure and the canonical efficiency factors of any generalized interaction can be calculated from the designs obtained by deleting all factors not in the interaction. For instance, the efficiency factors of the main effect of factor F_1 are given by those of the design with incidence matrix $\mathbf{N}_1$, while the efficiency factors of the main effects of F_1 and F_2 and of the $F_1 F_2$ interaction are given by the design with incidence matrix $\mathbf{N}_1 \otimes \mathbf{N}_2$. However, these n-factor designs have little practical value as the block size k or the number of replicates r or both will often be too large.

To overcome these difficulties Mukerjee (1981) and Gupta (1983) have generalized the method to produce a number of series of designs with smaller block sizes and replications. Their methods produce very general families of designs, including for instance n-cyclic designs, which possess the same properties in terms of factorial structure and efficiency factors as the designs produced by (9.30). In general, however, the designs cannot be constructed in a simple systematic manner nor are explicit expressions available for the calculation of efficiency factors. Obtaining a factorial design of a given size with certain desirable properties could be a daunting task.

9.8 Row–column designs

Factorial experiments can also be set out in $k \times s$ row–column designs as the example of Section 9.6.3 shows. The simplest type of design would be to use a Latin square for the v treatment combinations. The main effects and interactions would then be estimated independently of both row and column parameters. If insufficient experimental material was available to accommodate all treatment combinations in either the rows or the columns then a row-orthogonal row–column design could be used, see Section 5.7. For instance, the 4^2 design in Example 9.1 has been set out in such a way that each of the 16 treatment combinations occurs once in each position within the blocks; the resulting design is thus a Youden square. The design in Table 9.4 is a row-orthogonal row–column design for a 3×5 experiment in five rows and 15 columns.

In factorial experiments, however, the number of treatment combinations v is often too large to use either a Latin square or a row-orthogonal design. It will frequently be the case that it will be necessary for both the number of rows and columns to be less than v. Certain treatment comparisons will, therefore, be confounded with the row and column effects. Yates (1937) introduced quasi-Latin squares and demonstrated their use as row–column designs for factorial experiments. Both Yates (1937) and Rao (1946) gave numerous designs, mainly for factors at two or three levels, in which certain main effects and interactions were either totally or partially confounded with rows or columns. The

DSIGN algorithm given by Patterson (1976) is a general procedure for constructing factorial designs in various blocking structures including row–column designs. Examples of row–column factorial designs produced by this algorithm are given in Bailey (1977) and Patterson and Bailey (1978).

The n-cyclic designs defined in Section 3.5 have been shown to provide a flexible method of generating single, fractional and multiple replicate block designs for factorial experiments. John and Lewis (1983) have considered the use of such designs in row–column factorial experiments. Their method includes quasi-Latin squares and the row–column designs produced by the DSIGN algorithm if attention is again restricted to the subset of designs considered by Bailey (1977); see Section 8.11.

The method of construction ensures that the row–column design is such that both the row component design D_k and the column component design D_s are n-cyclic designs. Suppose the initial row consists of a subgroup G_1 of the treatment combinations of order d_1 together with a further $(s/d_1) - 1$ of its cosets. Since an n-cyclic set of v/d_1 blocks is generated from this row by the addition of one treatment combination from each of the cosets of G_1, the initial column must consist of kd_1/v elements from each of the cosets of G_1. A similar argument shows that the initial column must consist of a subgroup G_2 of order d_2 together with a further $(k/d_2) - 1$ of its cosets and that the initial row must consist of sd_2/v elements from each of the cosets of G_2. Clearly d_1 and d_2 must be common factors of v and s and v and k respectively, and v must be a divisor of kd_1 and sd_2. It can be easily verified that the resulting n-cyclic row–column design satisfies the adjusted orthogonality property given by (5.11) and has factorial structure.

Example 9.8
Consider the construction of a single replicate 3^4 experiment in a 9×9 row–column design. It is necessary to ensure that the initial row and column are chosen so that they only have treatment combination 0000 in common. This can be achieved, for instance, by taking the treatment combinations in the initial row to be given by the direct sum of the groups generated from 1011 and 1102, and in the initial column by the direct sum of groups generated from 1101 and 0112. If the methods of Section 8.7 are applied separately to the row and column component designs, it can be shown that $F_1 F_2^2 F_3^2$, $F_1 F_2 F_4^2$, $F_1 F_3 F_4$ and $F_2 F_3^2 F_4$ are confounded with rows and $F_1 F_2^2 F_3$, $F_1 F_2 F_4$, $F_1 F_3^2 F_4^2$ and $F_2 F_3 F_4^2$ with columns. The

full design is given by

0000	1011	2022	1102	2110	0121	2201	0212	1220
1101	2112	0120	2200	0211	1222	0002	1010	2021
2202	0210	1221	0001	1012	2020	1100	2111	0122
0112	1120	2101	1211	2222	0200	2010	0021	1002
1210	2221	0202	2012	0020	1001	0111	1122	2100
2011	0022	1000	0110	1121	2102	1212	2220	0201
0221	1202	2210	1020	2001	0012	2122	0100	1111
1022	2000	0011	2121	0102	1110	0220	1201	2212
2120	0101	1112	0222	1200	2211	1021	2002	0010

Example 9.9
John and Lewis (1983) give an example of a 3×4 experiment in a 4×6 row–column design. The initial row consists of subgroup (00 10 20) and coset (01 11 21), while the initial column consists of subgroup (00 02) and coset (11 13). The full design is

00	01	10	11	20	21
02	03	12	13	22	23
11	12	21	22	01	02
13	10	23	20	03	00

It can be verified that each row has two treatments in common with every column, and thus satisfies the adjusted orthogonality condition. The canonical efficiency factors of the design can be obtained from (9.25) using the elements from the information matrix given by (5.3). Alternatively, the canonical efficiency factors can be obtained for the row and column components separately using (9.25) and then combined using (5.12). The efficiency factors for the F_1 and F_2 main effects are 0.75, 0.75 and 0.5, 1, 0.5 respectively, and for the $F_1 F_2$ interaction 1, 0.25, 1, 1. Further examples are given in John and Lewis (1983).

It may also be desirable to set out factorial experiments in resolvable and Latinized row–column designs. For example, suppose a forestry trial is planned to determine the best combinations of eight different seedlots and four salt/irrigation levels. The experiment is to be set out in a glasshouse and, for the reasons discussed in Chapter 6, a two-dimensional blocking structure of rows and columns within replicates is considered appropriate. A design would then be required for a two-factor experiment using a number of replicates and where each replicate comprises, say, four rows and eight columns. Further, if the replicates are contiguous then a design satisfying the Latinized property of Section 6.5 would also be required.

The additional constraints imposed by these blocking factors makes it increasingly difficult to find flexible classes of appropriate designs that satisfy the requirement of factorial structure. That is, in order to construct designs using these more complex blocking structures it will be necessary to sacrifice the desirable property that main effects and interactions are orthogonal to each other.

Interchange algorithms to generate designs for a wide range of blocking structures have been discussed in detail in Section 6.4. These algorithms can also be used for different treatment structures, such as factorial experiments. The required features of the design will be determined by the choice of appropriate objective functions. Following the criteria used in Section 2.5 for choosing efficient designs, an intuitively appealing requirement for a factorial resolvable row–column design is to ensure that the levels of each factor occur as equally as possible in every row and in every column of each replicate. This requirement was used in the construction of n-cyclic factorial designs in Section 9.6.2. In this case it will ensure that the main effects are estimated with a high degree of precision. For two-factor interactions the requirement would apply to the combination of pairs of treatment factors, and so on for higher level interactions. Of course, it may not be possible to satisfy these requirements for all the factorial effects of interest, so that some compromise will be necessary.

Williams and John (1995) have proposed objective functions that aim to ensure the efficient estimation of main effects. Following from (2.15) and (5.18), an appropriate criterion for the ith main effect can be based on

$$t_i = \text{trace}(\mathbf{W}_i) + \text{trace}(\mathbf{W}_i^2) \tag{9.31}$$

where

$$\mathbf{W}_i = (1/r_i s)\mathbf{N}_{ki}\mathbf{N}_{ki}' + (1/r_i k)\mathbf{N}_{si}\mathbf{N}_{si}' - 2\mathbf{K}$$

and where $\mathbf{N}_{ki}$ and $\mathbf{N}_{si}$ are the incidence matrices of the row and column component main effect designs given by deleting all factors except the ith from the full resolvable row–column design. The first term on the right-hand side of (9.31) is required as the component designs will usually be non-binary.

The t_i $(i = 1, 2, \ldots, n)$ values obtained for each main effect design can now be combined into a single objective function. Before doing so, however, they have to be standardized so that they are numerically comparable. Williams and John (1995) suggested the use of a lower bound for t_i given when the resolvable row–column design is efficiency-balanced. One such bound is given by

$$L_i = (m_i - 1)(1 - U_i)(2 - U_i)$$

where U_i is the upper bound (5.15) for the average efficiency factor of the resolvable row–column design for the ith treatment factor. An overall objective function f is then given by the weighted average of the standardized values $f_i = L_i/t_i$ $(i = 1, 2, \ldots, n)$, where factor levels are used as weights. That is,

$$f = \frac{\sum_{i=1}^n m_i f_i}{\sum_{i=1}^n m_i} \tag{9.32}$$

Example 9.10
Using the simulated annealing algorithm of John and Whitaker (1993) with objective function f given in (9.32) produced the following two-factor design for $v = 12$, $n = 2$, $m_1 = 4$, $m_2 = 3$, $k = 4$, $s = 3$ and $r = 3$:

	Replicate	
1	2	3
10 22 31	02 11 20	21 02 30
20 11 02	30 21 12	00 11 22
01 30 12	10 32 01	10 32 01
32 00 21	22 00 31	31 20 12

It can be seen that levels of the first factor occur once in each column and at most once in each row, within replicates. The levels of the second factor occur once in each row and at most twice in each column, however, not all the levels of the factor appear in every column.

The algorithm can be used to produce single replicate resolvable row–column designs. In particular, it provides a means of constructing the two-factor contractions needed to give the two replicate row–column designs of Section 6.6.

Additional terms can be added to the objective function f in (9.32) to ensure that the overall resolvable row–column design is efficient and, if required, satisfies the Latinized property of Section 6.5.

Further research on the use of these methods to produce

good factorial designs for complex blocking structures is needed, especially to ensure that the interactions as well as the main effects are estimated as precisely as possible.

Some matrix results

The main matrix results required for this book are given in this appendix. Most of them are stated without proof as they are relatively straightforward and can be found in most books on matrix algebra; see for instance Searle (1966), Basilevsky (1983) and Graybill (1983). The properties of circulant matrices are considered in more detail as they have particular relevance in the book and are perhaps not widely known. A detailed account of circulant matrices can be found in Davis (1979).

A.1 Notation

An $n \times m$ matrix $\mathbf{A} = (a_{ij})$ has n rows and m columns with a_{ij} being the element in the ith row and jth column. $\mathbf{A}'$ will denote the transpose of $\mathbf{A}$. $\mathbf{A}$ is *symmetric* if $\mathbf{A} = \mathbf{A}'$. If $m = 1$ then the matrix will be called a *column vector* and will usually be denoted by a lower-case letter, $\mathbf{a}$ say. The transpose of $\mathbf{a}$ will be a *row vector* $\mathbf{a}'$.

A diagonal matrix whose diagonal elements are those of the vector $\mathbf{a}$, and whose off-diagonal elements are zero, will be denoted by $\mathbf{a}^{\delta}$. The inverse of $\mathbf{a}^{\delta}$ will be denoted by $\mathbf{a}^{-\delta}$. More generally, if each element in $\mathbf{a}$ is raised to power p then the corresponding diagonal matrix will be denoted by $\mathbf{a}^{p\delta}$.

Some special matrices and vectors are:

$$\begin{array}{ll}
\mathbf{1}_n, & \text{the } n \times 1 \text{ vector with every element unity} \\
\mathbf{I}_n = \mathbf{1}_n^{\delta}, & \text{the } n \times n \text{ identity matrix} \\
\mathbf{J}_{n \times m} = \mathbf{1}_n \mathbf{1}_m', & \text{the } n \times m \text{ matrix with every element unity} \\
\mathbf{J}_n = \mathbf{J}_{n \times n} & \\
\mathbf{K}_n = (1/n)\mathbf{J}_n, & \text{the } n \times n \text{ matrix with every element } n^{-1}.
\end{array}$$

Suffixes can be omitted if, in doing so, no ambiguity results.

A.2 Trace and rank

If $\mathbf{A} = (a_{ij})$ is a square matrix of order n, i.e. an $n \times n$ matrix, then

$$\text{trace}(\mathbf{A}) = \sum_{i=1}^{n} a_{ii}$$

The number of linearly independent rows or columns of the $n \times n$ matrix $\mathbf{A}$ is given by $r = \text{rank}(\mathbf{A})$. If $r = n$ then $\mathbf{A}$ is non-singular, if $r < n$ $\mathbf{A}$ is singular.

Provided the matrices are conformable, though not necessarily square, then

$$\text{trace}(\mathbf{AB}) \quad = \quad \text{trace}(\mathbf{BA}) \qquad\qquad\qquad (\text{A}.1)$$

$$\text{rank}(\mathbf{AB}) \quad \leq \quad \min\{\text{rank}(\mathbf{A}), \text{rank}(\mathbf{B})\} \qquad (\text{A}.2)$$

$$\text{rank}(\mathbf{AA}') \quad = \quad \text{rank}(\mathbf{A}'\mathbf{A}) = \text{rank}(\mathbf{A}) = \text{rank}(\mathbf{A}') \quad (\text{A}.3)$$

It follows from (A.1) that the trace of the product of matrices is invariant under any cyclic permutation of the matrices. For example

$$\text{trace}(\mathbf{ABC}) = \text{trace}(\mathbf{BCA}) = \text{trace}(\mathbf{CAB})$$

A.3 Eigenvalues and eigenvectors

Let $\mathbf{A}$ be a square symmetric matrix of order n. An *eigenvalue* of $\mathbf{A}$ is a scalar θ such that $\mathbf{Ax} = \theta\mathbf{x}$ for some vector $\mathbf{x} \neq \mathbf{0}$. The vector $\mathbf{x}$ is called an *eigenvector* of $\mathbf{A}$.

If $\theta_1, \theta_2, \ldots, \theta_n$ are the eigenvalues of $\mathbf{A}$ then

$$\text{trace}(\mathbf{A}) \quad = \quad \sum_{i=1}^{n} \theta_i \qquad\qquad\qquad\qquad (\text{A}.4)$$

$$\text{rank}(\mathbf{A}) \quad = \quad \text{number of non-zero eigenvalues} \qquad (\text{A}.5)$$

$$|\mathbf{A}| \quad = \quad \prod_{i=1}^{n} \theta_i \qquad\qquad\qquad\qquad (\text{A}.6)$$

where $|\mathbf{A}|$ is the determinant of $\mathbf{A}$. If $\theta_1, \theta_2, \ldots, \theta_m$ are the *distinct* eigenvalues of $\mathbf{A}$ ($m \leq n$) then the determinant can be written as

$$|\mathbf{A}| = \theta_1^{n_1} \theta_2^{n_2} \ldots \theta_m^{n_m}$$

where n_i is the *multiplicity* of θ_i, and where $\sum n_i = n$.

Repeated application of $\mathbf{Ax} = \theta\mathbf{x}$ shows that, for some positive integer h, $\mathbf{x}$ is also an eigenvector of $\mathbf{A}^h$ with corresponding eigenvalue θ^h, i.e.

$$\mathbf{A}^h\mathbf{x} = \theta^h\mathbf{x} \quad (h = 1, 2, \ldots) \tag{A.7}$$

The eigenvalues of a symmetric matrix are real. For each eigenvalue of a symmetric matrix there exists a real eigenvector.

If $\mathbf{B}$ is a non-square matrix then $\mathbf{B}'\mathbf{B}$ and $\mathbf{BB}'$ have the same non-zero eigenvalues.

For every symmetric matrix $\mathbf{A}$ there exists a non-singular matrix $\mathbf{X}$ such that

$$\mathbf{X}^{-1}\mathbf{AX} = \theta^\delta \tag{A.8}$$

where the elements of θ are the eigenvalues of $\mathbf{A}$.

For the symmetric matrix $\mathbf{A}$ there exists a set of orthogonal and normalized eigenvectors $\mathbf{x}_1, \mathbf{x}_2, \ldots, \mathbf{x}_n$ satisfying

$$\mathbf{x}_i'\mathbf{x}_j = \left\{ \begin{array}{ll} 1, & i = j \\ 0, & i \neq j \end{array} \right.$$

The *canonical* or *spectral* decomposition of $\mathbf{A}$ is then given by

$$\mathbf{A} = \sum_{i=1}^{n} \theta_i \mathbf{x}_i \mathbf{x}_i' \tag{A.9}$$

where θ_i is the eigenvalue corresponding to $\mathbf{x}_i$. If $\mathbf{A}$ is non-singular then

$$\mathbf{A}^{-1} = \sum_{i=1}^{n} \theta_i^{-1} \mathbf{x}_i \mathbf{x}_i' \tag{A.10}$$

It also follows from (A.9) that the eigenvectors $\mathbf{x}_i$ $(i = 1, 2, \ldots, n)$ satisfy

$$\sum_{i=1}^{n} \mathbf{x}_i \mathbf{x}_i' = \mathbf{I} \tag{A.11}$$

A.4 Idempotent matrices

A square matrix $\mathbf{A}$ is *idempotent* if $\mathbf{A}^2 = \mathbf{A}$. If $\mathbf{A}$ is idempotent then so is $\mathbf{I} - \mathbf{A}$. $\mathbf{I}$ and $\mathbf{K}$ are examples of idempotent matrices.

The eigenvalues of an idempotent matrix $\mathbf{A}$ of order n are 1 with multiplicity r and 0 with multiplicity $n - r$, where $r = \text{rank}(\mathbf{A})$. It then follows from (A.4) and (A.5) that for an idempotent matrix

$$\text{rank}(\mathbf{A}) = \text{trace}(\mathbf{A}) \tag{A.12}$$

If $\mathbf{A}$ and $\mathbf{B}$ are symmetric idempotent matrices then $\mathbf{AB}$ is idempotent if and only if $\mathbf{AB} = \mathbf{BA}$.

A.5 Generalized inverses

Let $\mathbf{A}$ be a square matrix of order n with $\text{rank}(\mathbf{A}) = r$. If $r = n$ then $\mathbf{A}$ has a unique inverse $\mathbf{A}^{-1}$ which satisfies

$$\mathbf{A}\mathbf{A}^{-1} = \mathbf{A}^{-1}\mathbf{A} = \mathbf{I} \tag{A.13}$$

Further, a unique solution to the equations $\mathbf{A}\mathbf{x} = \mathbf{y}$ is then given by

$$\mathbf{x} = \mathbf{A}^{-1}\mathbf{y} \tag{A.14}$$

If $r < n$, $\mathbf{A}^{-1}$ does not exist. There will, however, be an infinite number of solutions of the consistent equations $\mathbf{A}\mathbf{x} = \mathbf{y}$. As an example, consider the equations

$$\begin{pmatrix} 1 & 2 \\ 2 & 4 \end{pmatrix} \begin{pmatrix} x_1 \\ x_2 \end{pmatrix} = \begin{pmatrix} 3 \\ 6 \end{pmatrix}$$

The second equation is twice the first so that $\text{rank}(\mathbf{A}) = 1$. Replacing 6 by 5, say, in the vector $\mathbf{y}$ would lead to an inconsistent set of equations for which no solution will exist.

One solution to the above equations is given by $x_1 = 3$, $x_2 = 0$ and can be written as $\mathbf{x} = \mathbf{B}\mathbf{y}$ where

$$\mathbf{B} = \begin{pmatrix} 1 & 0 \\ 0 & 0 \end{pmatrix}$$

Although $\mathbf{A}^{-1}$ does not exist, the matrix $\mathbf{B}$ plays the role of an inverse in that its use leads to a solution to the equations. $\mathbf{B}$ is called a *generalized inverse* or *g-inverse* of $\mathbf{A}$, and is denoted by $\mathbf{A}^-$; it is sometimes called a pseudo inverse or conditional inverse of $\mathbf{A}$. It is not unique since

$$\mathbf{A}^- = \begin{pmatrix} 1 & 0 \\ 2 & -1 \end{pmatrix}$$

also leads to the solution $x_1 = 3$, $x_2 = 0$, whereas

$$\mathbf{A}^- = \begin{pmatrix} 1/3 & 0 \\ 0 & 1/6 \end{pmatrix}$$

leads to a different solution, namely $x_1 = x_2 = 1$.

$\mathbf{A}^-$ is a g-inverse of $\mathbf{A}$ if and only if it satisfies

$$\mathbf{A}\mathbf{A}^-\mathbf{A} = \mathbf{A} \tag{A.15}$$

A solution to the consistent equations $\mathbf{A}\mathbf{x} = \mathbf{y}$ is then given by

$$\mathbf{x} = \mathbf{A}^-\mathbf{y} \tag{A.16}$$

It can be verified that the three g-inverses given in the above example satisfy (A.15).

The matrix $\mathbf{A}^+$ which satisfies the following four conditions:

$$\begin{aligned}
\mathbf{A}\mathbf{A}^+\mathbf{A} &= \mathbf{A} \\
\mathbf{A}^+\mathbf{A}\mathbf{A}^+ &= \mathbf{A}^+ \\
(\mathbf{A}\mathbf{A}^+)' &= \mathbf{A}\mathbf{A}^+ \\
(\mathbf{A}^+\mathbf{A})' &= \mathbf{A}^+\mathbf{A}
\end{aligned} \tag{A.17}$$

is called the *Moore–Penrose generalized inverse* of $\mathbf{A}$; it can be shown to be a unique matrix.

If $\mathbf{A}$ is given in canonical form as

$$\mathbf{A} = \sum_{i=1}^n \theta_i \mathbf{x}_i \mathbf{x}_i'$$

where $\theta_1, \theta_2, \ldots, \theta_r$ $(r < n)$ are the non-zero eigenvalues, then

$$\mathbf{A}^+ = \sum_{i=1}^r \theta_i^{-1} \mathbf{x}_i \mathbf{x}_i' \tag{A.18}$$

is the Moore–Penrose generalized inverse of $\mathbf{A}$. More generally

$$\mathbf{A}^- = \mathbf{A}^+ + \sum_{i=r+1}^n \alpha_i \mathbf{x}_i \mathbf{x}_i' \tag{A.19}$$

is a g-inverse of $\mathbf{A}$ for a given set of constants α_i $(i = r + 1, r + 2, \ldots, n)$.

If $\mathbf{A}$ is a symmetric idempotent matrix then two useful g-inverses of $\mathbf{A}$ are the Moore–Penrose generalized inverse $\mathbf{A}^+ = \mathbf{A}$ and the inverse $\mathbf{A}^- = \mathbf{I}$.

The following results are frequently used in this book:

(i) $\mathbf{A}\mathbf{A}^-$ is an idempotent matrix with

$$\text{rank}(\mathbf{A}\mathbf{A}^-) = \text{trace}(\mathbf{A}\mathbf{A}^-) = \text{rank}(\mathbf{A}) \tag{A.20}$$

(ii) If $(\mathbf{A}'\mathbf{A})^-$ is a g-inverse of $\mathbf{A}'\mathbf{A}$ then

$$\mathbf{A}(\mathbf{A}'\mathbf{A})^-\mathbf{A}'\mathbf{A} = \mathbf{A} \tag{A.21}$$

(iii) If $\mathbf{A}$ is an $n \times n$ matrix with $\mathbf{A}\mathbf{1} = \mathbf{0}$ and rank $n - 1$, then it follows directly from (A.11) that

$$\mathbf{A}^+\mathbf{A} = \mathbf{I} - \mathbf{K} \tag{A.22}$$

A.6 Kronecker products

Let $\mathbf{A} = (a_{ij})$ be a $p \times q$ matrix and $\mathbf{B}$ be an $r \times s$ matrix. The *Kronecker product* of $\mathbf{A}$ and $\mathbf{B}$, denoted by $\mathbf{A} \otimes \mathbf{B}$, is the $pr \times qs$ matrix defined as

$$\mathbf{A} \otimes \mathbf{B} = \begin{pmatrix} a_{11}\mathbf{B} & a_{12}\mathbf{B} & \cdots & a_{1q}\mathbf{B} \\ a_{21}\mathbf{B} & a_{22}\mathbf{B} & \cdots & a_{2q}\mathbf{B} \\ \vdots & \vdots & \ddots & \vdots \\ a_{p1}\mathbf{B} & a_{p2}\mathbf{B} & \cdots & a_{pq}\mathbf{B} \end{pmatrix}$$

Two special results are that $\mathbf{I}_n \otimes \mathbf{I}_m = \mathbf{I}_{nm}$ and $\mathbf{K}_n \otimes \mathbf{K}_m = \mathbf{K}_{nm}$.

If a is a scalar and matrices are conformable, then

$$
\begin{aligned}
(a\mathbf{A}) \otimes \mathbf{B} &= \mathbf{A} \otimes (a\mathbf{B}) = a(\mathbf{A} \otimes \mathbf{B}) \\
(\mathbf{A} \otimes \mathbf{B}) \otimes \mathbf{C} &= \mathbf{A} \otimes (\mathbf{B} \otimes \mathbf{C}) \\
(\mathbf{A} \otimes \mathbf{B})' &= \mathbf{A}' \otimes \mathbf{B}' \\
(\mathbf{A} \otimes \mathbf{B})(\mathbf{C} \otimes \mathbf{D}) &= \mathbf{AC} \otimes \mathbf{BD} \\
(\mathbf{A} + \mathbf{B}) \otimes \mathbf{C} &= \mathbf{A} \otimes \mathbf{C} + \mathbf{B} \otimes \mathbf{C}
\end{aligned}
\tag{A.23}
$$

Let $\mathbf{A}$ and $\mathbf{B}$ be square matrices of order n and m respectively, then

$$
\begin{aligned}
(\mathbf{A} \otimes \mathbf{B})^- &= \mathbf{A}^- \otimes \mathbf{B}^- & (A.24) \\
\text{trace}(\mathbf{A} \otimes \mathbf{B}) &= \text{trace}(\mathbf{A}).\text{trace}(\mathbf{B}) & (A.25)
\end{aligned}
$$

If $\mathbf{x}_1$ is an eigenvector of $\mathbf{A}$ with eigenvalue θ_1, and $\mathbf{x}_2$ is an eigenvector of $\mathbf{B}$ with eigenvalue θ_2, then

$$(\mathbf{A} \otimes \mathbf{B})(\mathbf{x}_1 \otimes \mathbf{x}_2) = \theta_1\theta_2(\mathbf{x}_1 \otimes \mathbf{x}_2) \tag{A.26}$$

i.e. $\mathbf{x}_1 \otimes \mathbf{x}_2$ is an eigenvector of $\mathbf{A} \otimes \mathbf{B}$ with eigenvalue $\theta_1\theta_2$.

A.7 Circulant matrices

If $\mathbf{A} = (a_{ij})$ is an $n \times n$ matrix with $a_{ij} = a_{1m}$ where

$$m = \begin{cases} j - i + 1, & j \geq i \\ n - (j - i + 1), & j < i \end{cases}$$

then $\mathbf{A}$ is called a *circulant* matrix. Since the first row of a circulant matrix completely determines the matrix, the n elements in the first row of $\mathbf{A}$ will be denoted by $a_0, a_1, \ldots, a_{n-1}$. For example, the circulant matrix for $n = 4$ is

$$\mathbf{A} = \begin{pmatrix} a_0 & a_1 & a_2 & a_3 \\ a_3 & a_0 & a_1 & a_2 \\ a_2 & a_3 & a_0 & a_1 \\ a_1 & a_2 & a_3 & a_0 \end{pmatrix}$$

Let the $n \times n$ matrix $\boldsymbol{\Gamma}_h$ be a *basic* circulant matrix whose first row has 1 in the $(h+1)$th column and zero elsewhere. For instance, for $n = 4$,

$$\boldsymbol{\Gamma}_2 = \begin{pmatrix} 0 & 0 & 1 & 0 \\ 0 & 0 & 0 & 1 \\ 1 & 0 & 0 & 0 \\ 0 & 1 & 0 & 0 \end{pmatrix}$$

The general circulant matrix $\mathbf{A}$ can then be written as

$$\mathbf{A} = \sum_{h=0}^{n-1} a_h \boldsymbol{\Gamma}_h \tag{A.27}$$

The eigenvectors and eigenvalues of the circulant matrix $\mathbf{A}$ can be determined from those of $\boldsymbol{\Gamma}_1$, in view of (A.27) and the fact that

$$\boldsymbol{\Gamma}_h = \boldsymbol{\Gamma}_1^h \quad (h = 0, 1, \ldots, n-1) \tag{A.28}$$

Eigenvectors of $\boldsymbol{\Gamma}_1$ are given by

$$\boldsymbol{\gamma}_j = \frac{1}{\sqrt{(n)}} \begin{pmatrix} 1 \\ \omega^j \\ \omega^{2j} \\ \vdots \\ \omega^{(n-1)j} \end{pmatrix} \tag{A.29}$$

with corresponding eigenvalues

$$\theta_{1j} = \omega^j$$

for $j = 0, 1, \ldots, n-1$, where ω is given by

$$\omega = \exp(2\pi i/n) = \cos(2\pi/n) + i\sin(2\pi/n) \tag{A.30}$$

and where $i = \sqrt{(-1)}$. This result follows from the fact that $\omega^n = 1$.

Using (A.7) and (A.28), eigenvectors of $\boldsymbol{\Gamma}_h$ are then given by (A.29) with eigenvalues

$$\theta_{hj} = \omega^{jh}$$

for $h, j = 0, 1, \ldots, n-1$. Hence, eigenvectors of the general circulant matrix $\mathbf{A}$ are also given by (A.29) with eigenvalues

$$\theta_j = \sum_{h=0}^{n-1} a_h \omega^{jh} \tag{A.31}$$

for $j = 0, 1, \ldots, n - 1$. Note that the eigenvectors γ_j are independent of the elements of the matrix $\mathbf{A}$.

If $\mathbf{A}$ is symmetric then $a_h = a_{n-h}$ $(h = 1, 2, \ldots, m)$ so that

$$\theta_j = a_0 + \sum_{h=1}^{m} a_h^* \{\omega^{jh} + \omega^{j(n-h)}\} \tag{A.32}$$

where

$$m = \left\{ \begin{array}{ll} n/2, & n \text{ even} \\ (n-1)/2, & n \text{ odd} \end{array} \right.$$

and where

$$a_h^* = \left\{ \begin{array}{ll} a_h/2, & \text{if } n \text{ even and } h = n/2 \\ a_h, & \text{otherwise} \end{array} \right.$$

Now since

$$\sin\{2\pi j(n - h)/n\} = -\sin(2\pi jh/n)$$

it follows that

$$\omega^{jh} + \omega^{j(n-h)} = \cos(2\pi jh/n) + \cos\{2\pi j(n - h)/n\}$$

so that the eigenvalues of a symmetric circulant matrix $\mathbf{A}$ are given by

$$\theta_j = \sum_{h=0}^{n-1} a_h \cos(2\pi jh/n) \tag{A.33}$$

A spectral decomposition of $\mathbf{A}$ is given by

$$\mathbf{A} = \sum_{j=0}^{n-1} \theta_j \gamma_j \bar{\gamma}_j' \tag{A.34}$$

where $\bar{\gamma}_j$ is the *conjugate* of γ_j given by replacing ω in γ_j by its complex conjugate

$$\bar{\omega} = \exp(-2\pi i/n) = \cos(2\pi/n) - i\sin(2\pi/n)$$

Equation (A.34) can be verified by first showing that

$$\Gamma_1 = \sum_{j=0}^{n-1} \omega^j \gamma_j \bar{\gamma}_j'$$

The (k, l)th element of $\gamma_j \bar{\gamma}_j'$ is $(1/n)\omega^{(k-l)j}$, and since for $k > l$

$$\omega^{(k-l)j} = \omega^{-\{n-(k-l)\}j}$$

it follows that $\gamma_j \bar{\gamma}_j'$ is a circulant matrix, which can be written as

$$\gamma_j \bar{\gamma}_j' = \frac{1}{n} \sum_{h=0}^{n-1} \omega^{-hj} \Gamma_h \qquad (A.35)$$

Hence

$$\sum_{j=0}^{n-1} \omega^j \gamma_j \bar{\gamma}_j' = \frac{1}{n} \sum_h \sum_j \omega^{-(h-1)j} \Gamma_h = \Gamma_1$$

since

$$\sum_j \omega^{-(h-1)j} = \begin{cases} n, & h = 1 \\ 0, & h \neq 1 \end{cases}$$

Then

$$\mathbf{A} = \sum_h a_h \Gamma_1^h = \sum_j \left(\sum_h a_h \omega^{jh} \right) \gamma_j \bar{\gamma}_j'$$

which establishes (A.34). Note that

$$\gamma_j' \bar{\gamma}_k = \begin{cases} 1, & j = k \\ 0, & j \neq k \end{cases}$$

The Moore–Penrose generalized inverse $\mathbf{A}^+$ of $\mathbf{A}$ is also a circulant matrix since it can be written as

$$\mathbf{A}^+ = \sum \theta_j^{-1} \gamma_j \bar{\gamma}_j'$$

where the summation is over all r non-zero eigenvalues of $\mathbf{A}$, and where $r = \mathrm{rank}(\mathbf{A})$. If $r = n$ then $\mathbf{A}^+ = \mathbf{A}^{-1}$. Now using (A.35),

$$\mathbf{A}^+ = \sum_h \psi_h \Gamma_h \qquad (A.36)$$

where

$$\psi_h = (1/n) \sum_j \theta_j^{-1} \omega^{-hj}$$

If $\mathbf{A}$ is a symmetric matrix then, from (A.32), $\theta_j = \theta_{n-j}$ so that

$$\psi_h = (1/n) \sum_j \theta_j^{-1} \cos(2\pi jh/n) \qquad (A.37)$$

A.8 Block circulant matrices

Let Γ_{h_k} be a basic circulant of order n_k and let the $n \times n$ matrix Γ_h be defined by

$$\Gamma_h = \Gamma_{h_1} \otimes \Gamma_{h_2} \otimes \ldots \otimes \Gamma_{h_m}$$

where $n = n_1 n_2 \ldots n_m$. Then following (A.27) a general *block circulant* matrix $\mathbf{A}$ is defined by

$$\mathbf{A} = \sum_{h_1=0}^{n_1-1} \sum_{h_2=0}^{n_2-1} \cdots \sum_{h_m=0}^{n_m-1} a_{h_1 h_2 \ldots h_m} \Gamma_h \tag{A.38}$$

where $a_{h_1 h_2 \ldots h_m}$ is an element in the first row of $\mathbf{A}$. It follows from (A.26) that eigenvectors of $\mathbf{A}$ are given by

$$\gamma_{j_1} \otimes \gamma_{j_2} \otimes \ldots \otimes \gamma_{j_m} \tag{A.39}$$

where γ_j is given by (A.29), with corresponding eigenvalues

$$\theta_{j_1 j_2 \ldots j_m} = \sum_{h_1} \sum_{h_2} \cdots \sum_{h_m} a_{h_1 h_2 \ldots h_m} \omega^s$$

where

$$s = n \sum_{k=1}^{m} j_k h_k / n_k$$

for $j_k = 0, 1, \ldots, n-1$.

If $\mathbf{A}$ is symmetric then

$$\theta_{j_1 j_2 \ldots j_m} = \sum_{h_1} \cdots \sum_{h_m} a_{h_1 h_2 \ldots h_m} \cos \left\{ \sum_{k=1}^{m} (2\pi j_k h_k / n_k) \right\} \tag{A.40}$$

References

Agrawal, H. (1966) Some methods of construction of designs for two-way elimination of heterogeneity. *J. Amer. Statist. Assoc.*, **61**, 1153–1171.

Agrawal, H.L. and Prasad, J. (1982a) Some methods of construction of balanced incomplete block designs with nested rows and columns. *Biometrika*, **69**, 481–483.

Agrawal, H.L. and Prasad, J. (1982b) Some methods of construction of GD–RC and rectangular-RC designs. *Austral. J. Statist.*, **24**, 191–200.

Anderson, D.A. and Eccleston, J.A. (1985) On the construction of a class of efficient row–column designs. *J. Statist. Planning and Inference*, **11**, 131–134.

Bailey, R.A. (1977) Patterns of confounding in factorial designs. *Biometrika*, **64**, 597–603.

Bailey, R.A. (1989) Designs: mappings between structured sets. *Surveys in Combinatorics*, J.Siemons (ed.), 22–51, Cambridge University Press, Cambridge.

Bailey, R.A. (1991) Strata for randomized experiments. *J. Roy. Statist. Soc.*, B, **53**, 27–78.

Bailey, R.A. and Patterson, H.D. (1991) A note on the construction of row-and-column designs with two replicates. *J. Roy. Statist. Soc.*, B, **53**, 645–648.

Bailey, R.A., Gilchrist, F.H.L. and Patterson, H.D. (1977) Identification of effects and confounding patterns in factorial designs. *Biometrika*, **64**, 347–354.

Baird, D. and Mead, R. (1991) The empirical efficiency and validity of two neighbour models. *Biometrics*, **47**, 1473–1487.

Bartlett, M.S. (1978) Nearest neighbour models in the analysis of field experiments. *J. Roy. Statist. Soc.*, B, **40**, 147–174.

Basilevsky, A. (1983) *Applied Matrix Algebra in the Statistical Sciences*. North-Holland, New York.

Besag, J. and Kempton, R. (1986) Statistical analysis of field experiments using neighbouring plots. *Biometrics*, **42**, 231–251.

Bose, R.C. (1947) Mathematical theory of the symmetrical factorial design. *Sankhyā*, **8**, 107–166.

Bose, R.C. and Kishen, K. (1940) On the problem of confounding in general symmetrical factorial designs. *Sankhyā*, **5**, 21–36.

Bose, R.C. and Nair, K.R. (1939) Partially balanced incomplete block designs. *Sankhyā*, **4**, 337–372.

Box, G.E.P., Hunter, W.G. and Hunter, J.S. (1978) *Statistics for Experimenters*. J. Wiley & Sons, New York.

Butz, L. (1982) *Connectivity in Multi-Factor Designs*. Heldermann Verlag, Berlin.

Ceranka, B. and Mejza, S. (1979) On the efficiency factor for a contrast of treatment parameters. *Biom. J.*, **21**, 99–102.

Cheng, C.-S. (1986) A method for constructing balanced incomplete block designs with nested row and columns. *Biometrika*, **73**, 695–700.

Clatworthy, W.H. (1973) *Tables of Two-Associate-Class Partially Balanced Designs*. Nat. Bureau of Stand., Appl. Math. Series 63.

Cochran, W.G. and Cox, G.M. (1957) *Experimental Designs*, Second Edition. J. Wiley & Sons, New York.

Cotter, S.C. (1974) A general method of confounding for symmetrical factorial experiments. *J. Roy. Statist. Soc.*, B, **36**, 267–276.

Cox, D.R. (1958) *Planning of Experiments*. J. Wiley & Sons, New York.

Cullis, B.R. and Gleeson, A.C. (1991) Spatial analysis of field experiments – an extension to two dimensions. *Biometrics*, **47**, 1449–1460.

David, H.A. (1967) Resolvable cyclic designs. *Sankhyā*, A, **29**, 191–198.

Davis, P.J. (1979) *Circulant Matrices*. J. Wiley & Sons, New York.

de Hoog, F.R., Speed, T.P. and Williams, E.R. (1990) On a matrix identity associated with generalized least squares. *Linear Algebra Appl.*, **127**, 449–456.

Dean, A.M. and John, J.A. (1975) Single replicate factorial experiments in generalized cyclic designs: II. Asymmetrical arrangements. *J. Roy. Statist. Soc.*, B, **37**, 72–76.

Draper, N.R. and Smith, H. (1981) *Applied Regression Analysis*, Second Edition. J. Wiley & Sons, New York.

Eccleston, J.A. and Hedayat, A. (1974) On the theory of connected designs: characterization and optimality. *Ann. Statist.*, **2**, 1238–1255.

Eccleston, J.A. and Jones, B. (1980) Exchange and interchange procedures to search for optimal row-and-column designs. *J. Roy. Statist. Soc.*, B, **42**, 272–276.

Eccleston, J.A. and Kiefer, J. (1981) Relationships of optimality for individual factors of a design. *J. Statist. Planning and Inference*, **5**, 213–219.

Eccleston, J.A. and McGilchrist, C.A. (1985) Algebra of a row–column design. *J. Statist. Planning and Inference*, **12**, 305–310.

Federer, W.T. (1955) *Experimental Design*. Macmillan, New York.

Fisher, R.A. (1960) *The Design of Experiments*. Oliver and Boyd, Edinburgh.

Fisher, R.A. and Yates, F. (1963) *Statistical Tables for Biological, Agricultural and Medical Research*. Oliver and Boyd, Edinburgh.

Franklin, M.F. (1984) Constructing tables of minimum aberration p^{n-m} designs. *Technometrics*, **26**, 225–232.

Freeman, G.H. (1976a) On the selection of designs for comparative experiments. *Biometrics*, **32**, 195–199.

Freeman, G.H. (1976b) A cyclic method of constructing regular group divisible incomplete block designs. *Biometrika*, **63**, 555–558.

Fries, A. and Hunter, W.G. (1980) Minimum aberration 2^{k-p} designs. *Technometrics*, **22**, 601–608.

Gleeson, A.C. and Cullis, B.R. (1987) Residual maximum likelihood (REML) estimation of a neighbour model for field experiments. *Biometrics*, **43**, 277–288.

Graybill, F.A. (1983) *Matrices with Applications in Statistics*. Wadsworth, Belmont.

Greenfield, A.A. (1976) Selection of defining contrasts in two-level experiments. *Appl. Statist.*, **25**, 64–67.

Greenfield, A.A. (1978) Selection of defining contrasts in two-level experiments – a modification. *Appl. Statist.*, **27**, 78.

Gupta, S.C. (1983) Some new methods for constructing block designs having orthogonal factorial structure. *J. Roy. Statist. Soc.*, B, **45**, 297–307.

Hall, W.B. and Jarrett, R.G. (1981) Nonresolvable incomplete block designs with few replicates. *Biometrika*, **68**, 617–627.

Harshbarger, B. (1949) Triple rectangular lattices. *Biometrics*, **5**, 1–13.

Harshbarger, B. and Davis, L.L. (1952) Latinized rectangular lattices. *Biometrics*, **8**, 73–84.

Houtman, A.M. and Speed, T.P. (1983) Balance in designed experiments with orthogonal block structure. *Ann. Statist.*, **11**, 1069–1085.

Ipinyomi, R.A. and John, J.A. (1985) Nested generalized cyclic row–column designs. *Biometrika*, **72**, 403–409.

James, A.T. and Wilkinson, G.N. (1971) Factorization of the residual operator and canonical decomposition of nonorthogonal factors in the analysis of variance. *Biometrika*, **58**, 279–294.

Jansen, J., Douven, R.C.M.H. and van Berkum, E.E.M. (1992) An annealing algorithm for searching optimal block designs. *Biom. J.*, **34**, 529–538.

Jarrett, R.G. (1977) Bounds for the efficiency factor of block designs. *Biometrika*, **64**, 67–72.

Jarrett, R.G. (1983) Definitions and properties for m-concurrence designs. *J. Roy. Statist. Soc.*, B, **45**, 1–10.

Jarrett, R.G. (1989) A review of bounds for the efficiency factor of block designs. *Austral. J. Statist.*, **31**, 118–129.

Jarrett, R.G. and Hall, W.B. (1978) Generalized cyclic incomplete block designs. *Biometrika*, **65**, 397–401.

Jarrett, R.G. and Hall, W.B. (1982) Some design considerations for variety trials. *Utilitas Mathematica*, **21B**, 153–168.

John, J.A. (1965) A note on the analysis of incomplete block experiments. *Biometrika*, **52**, 633–636.

John, J.A. (1981) Efficient cyclic designs. *J. Roy. Statist. Soc.*, B, **43**, 76–80.

John, J.A. and Dean, A.M. (1975) Single replicate factorial experiments in generalized cyclic designs: I. Symmetrical arrangements. *J. Roy. Statist. Soc.*, B, **37**, 63–71.

John, J.A. and Eccleston, J.A. (1986) Row–column α-designs. *Biometrika*, **73**, 301–306.

John, J.A. and Lewis, S.M. (1983) Factorial experiments in generalized cyclic row–column designs. *J. Roy. Statist. Soc.*, B, **45**, 245–251.

John, J.A. and Quenouille, M.H. (1977) *Experiments: Design and Analysis*. Griffin, London.

John, J.A. and Smith, T.M.F. (1974) Sum of squares in non-full rank general linear hypotheses. *J. Roy. Statist. Soc.*, B, **36**, 107–108.

John, J.A. and Street, D.J. (1992) Bounds for the efficiency factor of row–column designs. *Biometrika*, **79**, 658–661.

John, J.A. and Street, D.J. (1993) Bounds for the efficiency factor of row–column designs (amendment). *Biometrika*, **80**, 712–713.

John, J.A. and Turner, G. (1977) Some new group divisible designs. *J. Statist. Planning and Inference*, **1**, 103–107.

John, J.A. and Whitaker, D. (1993) Construction of resolvable row–column designs using simulated annealing. *Austral. J. Statist.* **35**, 237–245.

John, J.A. and Williams, E.R. (1982) Conjectures for optimal block designs. *J. Roy. Statist. Soc.* B, **44**, 221–225.

John, J.A., Whitaker, D. and Triggs, C.M. (1993) Construction of cyclic designs using integer programming. *J. Statist. Planning and Inference*, **36**, 357–366.

John, J.A., Wolock, F.W. and David, H.A. (1972) *Cyclic Designs*. Nat. Bureau of Stand., Appl. Math. Series 62.

John, P.W.M. (1971) *Statistical Design and Analysis of Experiments*. Macmillan, New York.

Jones, B. (1976) An algorithm for deriving optimal block designs. *Technometrics*, **18**, 451–458.

Jones, B. (1979) Algorithms to search for optimal row-and-column design. *J. Roy. Statist. Soc.*, B, **41**, 210–216.

Jones, B. and Eccleston, J.A. (1980) Exchange and interchange procedures to search for optimal designs. *J. Roy. Statist. Soc.*, B, **42**, 238–243.

Kempthorne, O. (1947) A simple approach to confounding and fractional replication in factorial experiments. *Biometrika*, **34**, 255–272.

Kempton, R.A., Seraphin, J.C. and Sword, A.M. (1994) Statistical analysis of two dimensional variation in variety yield trials. *J. Agric. Sci.*, **122**, 335–342.

Kshirsagar, A.M. (1966) Balanced factorial designs. *J. Roy. Statist. Soc.*, B, **28**, 559–567.

Kurkjian, B. and Zelen, M. (1963) Applications of the calculus for factorial arrangements: I. Block and direct product designs. *Biometrika*, **50**, 63–73.

Lamacraft, R.R. and Hall, W.B. (1982) Tables of cyclic incomplete block designs: r = k. *Austral. J. Statist.*, **24**, 350–360.

Lewis, S.M. (1982) Generators for asymmetrical factorial experiments. *J. Statist. Planning and Inference*, **6**, 59–64.

Lewis, S.M. and Dean, A.M. (1985) A note on efficiency-consistent designs. *J. Roy. Statist. Soc.*, B, **47**, 261–262.

Lewis, S.M. and Dean, A.M. (1991) On general balance in row–column designs. *Biometrika*, **78**, 595–600.

McLean, R.A. and Anderson, V.L. (1984) *Applied Factorial and Fractional Designs.* Dekker, New York.

Mukerjee, R. (1979) Inter-effect-orthogonality in factorial experiments. *Calcutta Statist. Assoc. Bull.*, **28**, 83–108.

Mukerjee, R. (1981) Construction of effect-wise orthogonal factorial designs. *J. Statist. Planning and Inference*, **5**, 221–229.

National Bureau of Standards (1957) Fractional factorial experiment designs for factors at two levels. *Appl. Math. Series* 48.

National Bureau of Standards (1959) Fractional factorial experiment designs for factors at three levels. *Appl. Math. Series* 54.

National Bureau of Standards (1961) Fractional factorial designs for experiments with factors at two or three levels. *Appl. Math. Series* 58.

Nelder, J.A. (1965) The analysis of randomized experiments with orthogonal block structure: I. Block structure and the null analysis of variance; II. Treatment structure and the general analysis of variance. *Proc. Roy. Soc.*, A, **283**, 147–178.

Nelder, J.A. (1968) The combination of information in generally balanced designs. *J. Roy. Statist. Soc.*, B, **30**, 303–311.

Nguyen, N.-K. (1994) Construction of cyclic incomplete block designs by heuristic method. CSIRO-IAPP Biometrics Unit, Tech. Rep. 9402.

Nguyen, N.-K. and Williams, E.R. (1993a) An algorithm for constructing optimal resolvable row–column designs. *Austral. J. Statist.*, **35**, 363–370.

Nguyen, N.-K. and Williams, E.R. (1993b) Construction of row–column designs by computer. CSIRO-IAPP Biometrics Unit, Tech. Rep. 9353.

Papadakis, J.S. (1937) Méthode statistique pour des expériences sur champ. *Bull. de l'Institut d'Amélioration des Plantes à Salonique*, No. 23.

Park, D.K. and Dean, A.M. (1990) Average efficiency factors and adjusted orthogonality in multidimensional designs. *J. Roy. Statist. Soc.*, B, **52**, 361–368.

Paterson, L.J. (1983) Circuits and efficiency in incomplete block designs. *Biometrika*, **70**, 215–225.

Paterson, L.J. and Patterson, H.D. (1983) An algorithm for constructing α-lattice designs. *Ars Combinatoria*, **16A**, 87–98.

Patterson, H.D. (1965) The factorial combination of treatments in rotation experiments. *J. Agric. Sci.*, **65**, 171–182.

Patterson, H.D. (1976) Generation of factorial designs. *J. Roy. Statist. Soc.*, B, **38**, 175–179.

Patterson, H.D. and Bailey, R.A. (1978) Design keys for factorial experiments. *Appl. Statist.*, **27**, 335–343.

Patterson, H.D. and Hunter, E.A. (1983) The efficiency of incomplete block designs in National List and Recommended List cereal variety trials. *J. Agric. Sci.*, **101**, 427–433.

Patterson, H.D. and Robinson, D.L. (1989) Row-and-column designs with two replicates. *J. Agric. Sci.*, **112**, 73–77.

Patterson, H.D. and Silvey, V. (1980) Statutory and recommended list trials of crop varieties in the United Kingdom. *J. Roy. Statist. Soc.*, A, **143**, 219–252.

Patterson, H.D. and Thompson, R. (1971) Recovery of inter-block information when block sizes are unequal. *Biometrika*, **58**, 545–554.

Patterson, H.D. and Williams, E.R. (1976a) A new class of resolvable incomplete block designs. *Biometrika*, **63**, 83–92.

Patterson, H.D. and Williams, E.R. (1976b) Some theoretical results on general block designs. *Proc. 5th British Combin. Conf. Congressus Numerantium* **XV**, 489–496. Utilitas Mathematica, Winnipeg.

Patterson, H.D., Williams, E.R. and Hunter, E.A. (1978) Block designs for variety trials. *J. Agric. Sci.*, **90**, 395–400.

Pearce, S.C. (1963) The use and classification of non-orthogonal designs. *J. Roy. Statist. Soc.*, A, **126**, 353–377.

Pearce, S.C. (1974) Optimality of design in plot experiments. *Proc. 8th Int. Biometric Conf.*, Constanta, Romania.

Pearce, S.C. (1975) Row-and-column designs. *Appl. Statist.*, **24**, 60–74.

Pearce, S.C., Calinski, T. and Marshall, T.F. de C. (1974) The basic contrasts of an experimental design with special reference to the analysis of data. *Biometrika*, **61**, 449–460.

Preece, D.A. (1967) Nested balance incomplete block designs. *Biometrika*, **54**, 479–486.

Preece, D.A. (1994) Double Youden rectangles – an update with examples of size 5×11. *Discrete Math.*, **125**, 309–318.

Raghavarao, D. (1971) *Constructions and Combinatorial Problems in Design of Experiments*. J. Wiley & Sons, New York.

Raghavarao, D. and Federer, W.T. (1975) On connectedness in two-way elimination of heterogeneity designs. *Ann. Statist.*, **3**, 730–735.

Raghavarao, D. and Shah, K.R. (1980) A class of D_0 designs for two-way elimination of heterogeneity. *Commun. Statist. Theor. Meth.*, **A9**, 75–80.

Raktoe, B.L. (1969) Combining elements from distinct finite fields in mixed factorials. *Ann. Math. Statist.*, **40**, 498–504.

Rao, C.R. (1946) Confounded factorial designs in quasi-Latin squares. *Sankhyā*, **7**, 295–304.

Rice, C.G. and Izumi, K. (1984) Annoyance due to combinations of noises. *Proc. Inst. Acoustics*, 287–294.

Russell, K.G. (1976) The connectedness and optimality of a class of row–column designs. *Commun. Statist. Theory. Meth.*, **A5**, 1479–1488.

Russell, K.G. (1980) Further results on the connectedness and optimality of designs of type O:XB. *Commun. Statist. Theory. Meth.*, **A9**, 439–447.

Russell, K.G., Eccleston, J.A. and Knudsen, G.J. (1981) Algorithms for the construction of (M,S)-optimal block designs and row–column designs. *J. Statist. Comput. Simul.*, **12**, 93–105.

Searle, S.R. (1966) *Matrix Algebra for the Biological Sciences.* J. Wiley & Sons, New York.

Searle, S.R. (1971) *Linear Models.* J. Wiley & Sons, New York.

Shah, B.V. (1959) A generalization of partially balanced incomplete block designs. *Ann. Math. Statist.*, **30**, 1041–1050.

Shah, K.R. (1960) Optimality criteria for incomplete block designs. *Ann. Math. Statist.*, **31**, 791–794.

Shrikhande, S.S. (1951) Designs for two-way elimination of heterogeneity. *Ann. Math. Statist.*, **22**, 235–247.

Sihota, S.S. and Banerjee, K.S. (1981) On the algebraic structures in the construction of confounding plans in mixed factorial designs on the lines of White and Hultquist. *J. Amer. Statist. Assoc.*, **76**, 996–1001.

Singh, M. and Dey, A. (1979) Block designs with nested rows and columns. *Biometrika*, **66**, 321–326.

Sreenath, P.R. (1989) Construction of some balanced incomplete block designs with nested rows and columns. *Biometrika*, **76**, 399–402.

Street, D.J. (1981) Graeco-Latin and nested row and column designs. *Combinatorial Mathematics*, **8**, 304–313. Lecture Notes in Mathematics 884, Ed. K.L. McAvaney, Springer-Verlag, Berlin.

Tjur, T. (1990) A new upper bound for the efficiency factor of a block design. *Austral. J. Statist.*, **32**, 231–237.

Uddin, N. and Morgan, J.P. (1990) Some constructions for balanced incomplete block designs with nested rows and columns. *Biometrika*, **77**, 193–202.

Uddin, N. and Morgan, J.P. (1991) Two constructions for balanced incomplete block designs with nested rows and columns. *Statistica Sinica*, **1**, 229–232.

Venables, W.N. and Eccleston, J.A. (1993) Randomized search strategies for finding near-optimal block or row–column designs. *Austral. J. Statist.* **35**, 371–382.

Voss, D.T. and Dean, A.M. (1987) A comparison of classes of single replicate factorial designs. *Ann. Statist.*, **15**, 376–384.

White, D. and Hultquist, R.A. (1965) Construction of confounding plans for mixed factorial designs. *Ann. Math. Statist.*, **36**, 1256–1271.

Wilkinson, G.N., Eckert, S.R., Hancock, T.W. and Mayo, O. (1983) Nearest neighbour (NN) analysis of field experiments. *J. Roy. Statist. Soc.*, B, **45**, 151–211.

Williams, E.R. (1975a) A new class of resolvable block designs. PhD Thesis, University of Edinburgh.

Williams, E.R. (1975b) Efficiency-balanced designs. *Biometrika*, **62**, 686–689.

Williams, E.R. (1976) Resolvable paired-comparison designs. *J. Roy. Statist. Soc.*, B, **38**, 171–174.

Williams, E.R. (1977) A note on rectangular lattice designs. *Biometrics*, **33**, 411–414.

Williams, E.R. (1986a) Row and column designs with contiguous replicates. *Austral. J. Statist.*, **28**, 154–163.

Williams, E.R. (1986b) A neighbour model for field experiments. *Biometrika*, **73**, 279–287.

Williams, E.R. and John, J.A. (1993) Upper bounds for Latinized designs. *Austral. J. Statist.*, **35**, 229–236.

Williams, E.R. and John, J.A. (1995) Factorial experiments in row–column designs. *Appl. Statist.* To appear.

Williams, E.R. and Matheson, A.C. (1994) *Experimental Design and Analysis for Use in Tree Improvement*. CSIRO, Australia.

Williams, E.R. and Patterson, H.D. (1977) Upper bounds for efficiency factors in block designs. *Austral. J. Statist.*, **19**, 194–201.

Williams, E.R. and Talbot, M. (1993) *ALPHA+: Experimental Designs for Variety Trials*. Design User Manual. CSIRO, Canberra and SASS, Edinburgh.

Williams, E.R, Patterson, H.D. and John, J.A. (1976) Resolvable designs with two replications. *J. Roy. Statist. Soc.*, B, **38**, 296–301.

Williams, E.R., Patterson, H.D. and John, J.A. (1977) Efficient two-replicate resolvable designs. *Biometrics*, **33**, 713–717.

Williams, E.R., Ratcliff, D. and van Ewijk, P.H. (1986) The analysis of lattice square designs. *J. Roy. Statist. Soc.*, B, **48**, 314–321.

Worthley, R. and Banerjee, K.S. (1974) A general approach to confounding plans in mixed factorial experiments when the number of levels of a factor is any positive integer. *Ann. Statist.*, **2**, 579–585.

Yates, F. (1936a) Incomplete randomized designs. *Ann. Eugenics*, **7**, 121–140.

Yates, F. (1936b) A new method of arranging variety trials involving a large number of varieties. *J. Agric. Sci.*, **26**, 424–455.

Yates, F. (1937) A further note on the arrangement of variety trials: Quasi-Latin squares. *Ann. Eugenics*, **7**, 319–331.

Yates, F. (1939) The recovery of inter-block information in variety trials arranged in three-dimensional lattices. *Ann. Eugenics*, **9**, 136–156.

Yates, F. (1940a) Lattice squares. *J. Agric. Sci.*, **30**, 672–687.

Yates, F. (1940b) The recovery of inter-block information in balanced incomplete block designs. *Ann. Eugenics*, **10**, 317–325.

Author index

Subject index